CHEMICAL ECOLOGY

Wageningen UR Frontis Series

VOLUME 16

Series editor:

R.J. Bogers
*Frontis – Wageningen International Nucleus for Strategic Expertise,
Wageningen University and Research Centre, The Netherlands*

Online version at http://www.wur.nl/frontis

The titles published in this series are listed at the end of this volume

CHEMICAL ECOLOGY
From Gene to Ecosystem

Edited by

MARCEL DICKE
Laboratory of Entomology,
Wageningen University and Research Centre,
Wageningen, The Netherlands

and

WILLEM TAKKEN
Laboratory of Entomology,
Wageningen University and Research Centre,
Wageningen, The Netherlands

Springer

A C.I.P. Catalogue record for this book is available from the Library of Congress.

ISBN-10 1-4020-4783-5 (HB)
ISBN-13 978-1-4020-4783-1 (HB)
ISBN-10 1-4020-4792-4 (PB)
ISBN-13 978-1-4020-4792-3 (PB)

Published by Springer,
P.O. Box 17, 3300 AA Dordrecht, The Netherlands.

www.springer.com

Printed on acid-free paper

Printed in the Netherlands.

CONTENTS

PREFACE

In March 2005 a Spring School on "Chemical ecology – from gene to ecosystem" was held in Wageningen, organized by members of the Dutch Graduate Schools *Production Ecology and Resource Conservation* and *Experimental Plant Sciences* in collaboration with Frontis – Wageningen International Nucleus for Strategic Expertise. The aim of this Spring School was to bring together scientists and PhD students who are active in the field of chemical ecology at different levels of integration.

The field of chemical ecology has rapidly expanded in the past ten years. Traditionally, chemical ecologists have been active at the individual level, but this has expanded towards both the genome level and the ecosystem level, as well as to an integrated approach from the genome to the ecosystem level. Novel developments in molecular biology provide chemical ecologists with exciting new tools that allow to address old questions in unprecedented ways. The present book reflects the ideas that were presented during the Spring School. The research systems range from aquatic to terrestrial systems, involving plants and animals.

The students who participated in the Spring School each presented their own research, which was intensively discussed with the invited speakers and other participants of the Spring School. The event was an intensive meeting with well-appreciated exchange of ideas between participants who are active at different levels of integration. It was clear from the evaluation that the discussions were valuable to all participants. This book provides many ideas for future research in chemical ecology through a multidisciplinary approach.

Each chapter of this book has been subjected to peer review. The referees who have contributed to this peer review were: Teris A. van Beek, Harro J. Bouwmeester, Marien de Bruyne, John Carlson, Nicole van Dam, Jeff A. Harvey, André Kessler, Hans Helder, Willem Jan de Kogel, Joop J.A. van Loon, John. A. Pickett, Hans M. Smid, Junji Takabayashi, Ralph Tollrian and Felix L. Wäckers. Their comments have been much appreciated and have been a great help in further improving the contributions to this book. All chapters have been reviewed by the editors of this volume as well.

The organizing committee of the Spring School consisted of the following scientists of Wageningen University and Research Centre: Teris A. van Beek, Robert J. Bogers, Harro J. Bouwmeester, Frans Griepink, Willem Jan de Kogel, Joop J.A. van Loon, Willem Takken, Claudius van den Vijver and Marcel Dicke (chair).

We would like to thank Rob Bogers and Paulien van Vredendaal for their help in the editing and lay-out process, respectively, resulting in the production of this volume.

The editors,

Marcel Dicke
Willem Takken

Wageningen, September 2005

CHAPTER 1

CHEMICAL ECOLOGY

A multidisciplinary approach

WILLEM TAKKEN AND MARCEL DICKE

Laboratory of Entomology, Wageningen University and Research Centre, PO Box 8031, 6700 EH Wageningen, The Netherlands. E-mail: willem.takken@wur.nl

Abstract. Chemical information conveyance is an important phenomenon in the biology of plants and animals. This involves intraspecific chemical communication and its exploitation by heterospecific organisms. As a result food webs are overlaid with information webs that can have important consequences for community processes. A vast amount of research shows that both the emission of chemical information and the responses to it are often genetically controlled, and mediated by numerous interactions between an individual and its environment. Overall, it is argued that ecosystem functioning is much dependent on the responses of various community members to chemical cues, and that therefore knowledge on the chemical communication, from the genetic level to the ecosystem, is critical for our understanding of the functioning of populations, communities and ecosystems.
Keywords: gene; species; population; community; ecosystem; chemical communication

INTRODUCTION

Chemical ecology is the science that addresses the role of chemical cues in the interaction of organisms with their environment. One of the earliest and best-known examples of chemical communication is the use of sex pheromones by insects such as the silk moth *Bombyx mori.* The sex pheromone travels over great distances and attracts male conspecifics (Karlson and Butenandt 1959). Such intra-specific communication is one of the most widespread methods of chemical communication that occurs in all classes of the animal kingdom where sexual reproduction is the main avenue of reproduction. Indeed, sex pheromones have been described in all phyla of animals, including arthropods, fish, birds and mammals (Stoddart 1990). In addition to this example of chemical communication that mediates reproduction, numerous other types of intra-specific chemical communication exist as well, such as those involved in aggregation (Borden 1985; Wertheim et al. 2005), trail-marking (Traniello and Robson 1995) and defence (Brand et al. 1989). Whereas at first attention was paid mostly to the identification of the chemical cues concerned and to the direct behavioural effects these cues elicited, it was rapidly understood how

M. Dicke and W. Takken (eds.), Chemical Ecology: From Gene to Ecosystem, 1-8.

important these cues are in the ecology of species. Cues released into the environment are secreted onto a surface area (e.g., scent marking) or as volatiles that travel through space. On their journey to the intended receiver, cues pass through a highly variable environment affected by wind, temperature, moisture and physical obstructions such as plants, animals and rocks. The widespread use of chemical cues to communicate with conspecifics is indicative of the many advantages of this way of information conveyance.

Whilst pheromone communication is a highly important but still relatively limited aspect of chemical ecology, matters become much more complicated when inter-specific interactions are considered. After all, any chemical that is disseminated into the environment may be exploited by any other organism in the environment. Pheromones can be exploited by organisms from other species such as predators or parasitoids (Dicke and Sabelis 1992; Stowe et al. 1995). Moreover, apart from pheromones also other cues can mediate inter-specific interactions. Rudolfs (1922) described how chemical cues from mammals affected the behaviour of mosquitoes, attracting them from a distance. Mosquitoes such as some anopheline species recognize mammalian odours and use these to locate a blood source for food (Takken and Knols 1999). Apart from mammalian (or vertebrate) blood, mosquitoes also feed on plant sugars, and they recognize plant volatiles as well (Thornsteinson and Brust 1962; Healy and Jepson 1988; Foster and Hancock 1994). In higher animals such as reptiles and mammals, olfaction is common in foraging for food. Animals may detect food by smell and many predators locate their prey by chemical cues (Albone 1984; Ylonen et al. 2003). Herbivorous insects generally recognize their food by volatile and non-volatile cues produced by the plant (Visser 1986; Schoonhoven et al. in press). Such cues not only serve as attractants but can also act as repellents or arrestants. As with sex pheromones, early studies on the role of plant volatiles in animal behaviour focused on the identification of the chemicals and bioassay studies that showed their role in plant–herbivore interactions. Since then, a more complex picture of these interactions has emerged.

The chemical interactions discussed above mostly concern bitrophic interactions. More recently, the importance of chemical cues was investigated in multitrophic interactions. For example, the production of plant volatiles following herbivore attack can result in the attraction of carnivorous insects that kill the plant's enemies (Vet and Dicke 1992; Turlings et al. 1995). This was a first step towards appreciating the involvement of chemical information in the ecological context of food webs. Apart from the exclusive involvement of macro-organisms, micro-organisms may also be involved in chemical information conveyance. For instance, microbial organisms were found to affect the odour emission from human sweat that attracts blood-feeding mosquitoes (Braks et al. 2000), and malaria parasites influence the attraction of mosquito vectors so that they can be transmitted to other hosts (Lacroix et al. 2005). All together, these examples show that the interactions of an organism with its environment can be profoundly affected by chemical information conveyance.

Analytical chemistry

The nature of chemical information conveyance mandates that further understanding of the interactions is based upon knowledge of the chemicals involved. These can range from highly volatile to non-volatile compounds. Organisms can produce a vast diversity of chemicals often in small amounts, and modern technology allows for their identification by standard methodology, usually gas chromatography in combination with mass spectrometry (GC-MS) or HPLC and, recently, rapid developments occur in the development of novel, large-scale, metabolomic analytical methods (Fiehn 2002). However, the availability of an analytical chemical profile of the cues produced by an organism does not automatically lead to the discovery of the active compound(s). This requires extensive research including, e.g., sensory physiological and behavioural methods; examples of such research are presented in several chapters of this volume.

Molecular genetics

Rapid developments in the field of genomics result in a fast accumulation of sequenced genomes of various organisms, including *Caenorhabditis elegans, Drosophila melanogaster, Anopheles gambiae, Arabidopsis thaliana, Oryza sativa* and many others (Holt et al. 2002; Adams et al. 2000; Hodgkin et al. 1995). For some of these organisms, such as *Drosophila melanogaster* and *Arabidopsis thaliana,* large numbers of mutants are available in stock centres. These mutants, which may be altered in the production or perception of chemical cues, provide exciting tools to investigate the role of certain genes in chemically mediated interactions. Genes involved in olfaction have been identified in *Drosophila melanogaster* and *Anopheles gambiae.* By the silencing of a gene that is essential in the signal-transduction path of a pheromonal interaction, the organism may no longer be able to respond to the signal and suffer a significant disadvantage (Giarratani and Vosshall 2003). It was found that in mosquitoes the behavioural inhibition following a blood meal is accompanied by down-regulation of olfactory receptor genes on the antennae (Takken et al. 2001; Fox et al. 2001). As the number of fully sequenced genomes is rapidly expanding, it is becoming clear that there is considerable homology in olfactory receptor genes among animal species (Jacquin and Merlin 2004; Robertson et al. 2003; Vosshall 2003). This is likely to further enhance our knowledge about the genetic regulation of chemical communication.

Information on genome sequences may also allow for other manipulative experiments such as the specific down-regulation of certain genes through antisense or RNA-interference techniques (Kessler et al. 2004; Dicke et al. 2004). Therefore, the rapidly expanding knowledge of molecular genetics will provide exciting new tools for ecologists to investigate the function of genes in ecological interactions. This volume presents several of these developments.

Cellular regulation of cue production and perception

The production of chemical signals by animals and plants is regulated through hormones and signal transduction pathways (Jurenka 1996; Rafaeli 2005; Dicke and Van Poecke 2002). Animals perceive chemical cues in specialized organs where the cues bind to receptors, setting a cascade of signal transduction into motion. This is best known from studies in the nematode *Caenorhabditis elegans*, the frog *Xenopus laevis* and the fruitfly *Drosophila melanogaster*, but it is believed that the mechanism of chemical signal transduction is much the same, at least, throughout the animal kingdom (Hildebrand and Shepherd 1997; Dobritsa et al. 2003; Restrepo 2004). Examples of the cellular response to chemical cues in these model organisms may therefore serve as a starting point for investigating this process across a wide range of species. Plants may also perceive chemical cues from their environment, but specialized organs involved have not been reported. How plants perceive chemical signals, however, remains poorly known to date (Dicke and Bruin 2001; Baldwin et al. 2002).

Behavioural responses

We would not know of the existence of chemical communication without having studied the responses of plants and animals (Cardé and Bell 1995; Dicke and Bruin 2001). In animals we can observe typical behavioural responses such as movement towards or away from the chemical cues or a change in behaviour towards subsequent activities. An example of this latter is the observation that female pigs become receptive to boars when exposed to male odour, which is exploited in artificial-insemination methods in pigs through the application of the boar's pheromone (Gower et al. 1981). It has appeared that many behavioural activities of organisms are mediated by chemical cues. Simple responses to a specific cue have been described, but it is much more common for organisms to respond to a complex blend of chemical cues, sometimes derived from more than one species (e.g., Reddy and Guerrero 2004). Individual components may mediate different behavioural components that together constitute a complex behavioural response (Cardé and Minks 1997). Behavioural responses may be fixed and predictable (e.g., a response to a mate), or phenotypically plastic and subject to learning (e.g., responses to resources that are variable) (McCall and Kelly 2002). A single chemical cue may elicit responses in many different organisms in the environment and, thus, community members are linked in reticulate information webs.

Populations and communities

Animal and plant populations are composed of individuals that each produce chemical information and respond to cues. Population and community processes are not only influenced by direct effects of interactions such as mating, predation and defence but also by behavioural responses to chemical cues. These responses influence spatial distribution and interaction with community members. Therefore, the production of chemical cues and the responses to them are expressions of the phenotype that contribute to processes at the population and community level (Vet 1996; Kessler et al. 2004; Dicke et al. 2004). Individual behaviour can be fixed or phenotypically plastic. Phenotypic plasticity allows individuals to adjust their responses to current environmental conditions and may have important ecological consequences for species interactions and community processes (Agrawal 2001). Investigating the effects of an individual chemical cue on population and community processes may be carried out by comparatively investigating the effects of variation in the expression of a single gene in an otherwise similar genetic background. To do so, knowledge of the mechanisms of gene expression and gene function is essential. This will be highlighted in this volume.

Ecosystems

Processes at the ecosystem level include, for instance, spatial and temporal variation and dispersal of populations and individuals. Ecosystems are spatially complex and organisms with high dispersal capabilities may drive biodiversity patterns and ecosystem functions (Tscharntke et al. 2005). Dispersal capabilities are dependent on physical properties such as speed and mode of displacement, but also on the ability to perceive chemical information. Therefore, chemical cues are likely to influence ecosystem processes as well. This receives attention in several chapters in this volume.

Chemical ecology from gene to ecosystem

From the examples discussed above it is clear that chemical cues that mediate interactions between individuals influence processes at various levels of biological integration. Chemical ecology has often addressed mechanisms but recently interest in the effects of chemical signalling on community and ecosystem processes is rapidly increasing (Thaler 2002; Van Zandt and Agrawal 2004; Kessler et al. 2004; Dicke et al. 2004). To bring these issues together in one volume, we have asked leaders in the field of chemical ecology to discuss their views on this matter from their specific field of expertise. Scientists from as wide a field as molecular genetics to ecosystem analysis have contributed to present a general overview of the cutting-edge knowledge on this topic.

Chemical information conveyance plays a role in the biology of virtually all species. Chemical communication between conspecific individuals may be exploited by individuals of other species and the information web may have consequences for community and ecosystem processes in terrestrial and aquatic systems. In this volume examples are discussed of advances in our understanding of the role of chemical signalling in simple and complex systems as illustrated by recent research on plants (below and above ground) and animals. The cellular aspects of chemical ecology are illustrated by research on parasitic weeds and on members of two different insect orders (Diptera and Hymenoptera) and on nematodes. These examples serve as an illustration of how such processes are regulated and organized and show that there is a good deal of similarity among different species. The relevance of chemical ecology for community ecology is discussed for, e.g., the native tobacco plant *Nicotiana attenuata.* It is clear that chemical communication cannot be viewed from the perspective of a single species but should be placed in a multi-species context. Chemical information mediates interactions in communities and ecosystems. This is presented for plant–plant interactions, plant–animal interactions and inter-specific interactions of aquatic organisms. Finally, the implications of these findings are discussed with a view to the relevance of chemical communication for ecosystem functioning. The main take-home message of this volume is that in order to fully appreciate the influence of chemical signalling on community and ecosystem processes one needs thorough knowledge of the mechanisms of chemical information conveyance from the gene to the individual. Therefore, the highly multidisciplinary approach of modern chemical ecology is likely to make an important contribution to biology in the 21st century.

REFERENCES

Adams, M.D., Celniker, S.E., Holt, R.A., et al. 2000. The genome sequence of *Drosophila melanogaster*. *Science,* 287 (5461), 2185-2195.

Agrawal, A.A., 2001. Phenotypic plasticity in the interactions and evolution of species. *Science,* 294 (5541), 321-326.

Albone, E.S., 1984. *Mammalian semiochemistry: the investigation of chemical signals between mammals*. Wiley, Chichester.

Baldwin, I.T., Kessler, A. and Halitschke, R., 2002. Volatile signaling in plant-plant-herbivore interactions: what is real? *Current Opinion in Plant Biology,* 5 (4), 351-354.

Borden, J.H., 1985. Aggregation pheromones. *In:* Kerkut, G.A. and Gilbert, L.I. eds. *Comprehensive insect physiology biochemistry and pharmacology. Vol. 9. Behaviour*. Pergamon, Oxford, 257-285.

Braks, M.A.H., Scholte, E.J., Takken, W., et al. 2000. Microbial growth enhances the attractiveness of human sweat for the malaria mosquito, *Anopheles gambiae sensu stricto* (Diptera: Culicidae). *Chemoecology,* 10 (3), 129-134.

Brand, J.M., Page, H.M. and Lindner, W.A., 1989. Are ant alarm-defense secretions only for alarm defense? *Naturwissenschaften,* 6 (277).

Cardé, R.T. and Bell, W.J. (eds.), 1995. *Chemical ecology of insects. Vol. 2*. Chapman & Hall, New York.

Cardé, R.T. and Minks, A.K. (eds.), 1997. *Insect pheromone research: new directions*. Chapman & Hall, New York.

Dicke, M. and Bruin, J., 2001. Chemical information transfer between plants: back to the future. *Biochemical Systematics and Ecology,* 29 (10), 981-994.

Dicke, M. and Sabelis, M.W., 1992. Costs and benefits of chemical information conveyance: proximate and ultimate factors. *In:* Roitberg, B.D. and Isman, M.B. eds. *Insect chemical ecology: an evolutionary approach*. Chapman & Hall, New York, 122-155.

Dicke, M., Van Loon, J.J.A. and De Jong, P.W., 2004. Ecogenomics benefits community ecology. *Science,* 305 (5684), 618-619.

Dicke, M. and Van Poecke, R.M.P., 2002. Signalling in plant-insect interactions: signal transduction in direct and indirect plant defence. *In:* Scheel, D. and Wasternack, C. eds. *Plant signal transduction.* Oxford University Press, Oxford, 289-316. Frontiers in Molecular Biology No. 38.

Dobritsa, A.A., Van der Goes - Van Naters, W., Warr, C.G., et al. 2003. Integrating the molecular and cellular basis of odor coding in the *Drosophila* antenna. *Neuron,* 37 (5), 827-841.

Fiehn, O., 2002. Metabolomics: the link between genotypes and phenotypes. *Plant Molecular Biology,* 48 (1/2), 155-175.

Foster, W.A. and Hancock, R.G., 1994. Nectar-related olfactory and visual attractants for mosquitoes. *Journal of the American Mosquito Control Association,* 10 (2, part 2), 288-296.

Fox, A.N., Pitts, R.J., Robertson, H.M., et al. 2001. Candidate odorant receptors from the malaria vector mosquito *Anopheles gambiae* and evidence of down-regulation in response to blood feeding. *Proceedings of the National Academy of Sciences of the United States of America,* 98 (25), 14693-14697.

Giarratani, L. and Vosshall, L.B., 2003. Toward a molecular description of pheromone perception. *Neuron,* 39 (6), 881-883.

Gower, D.B., Hancock, M.R. and Bannister, L.H., 1981. Biochemical studies on the boar pheromones, 5-Androst-16-en-3-one and 5-Androst-16-en 3_-ol, and their metabolism by olfactory tissue. *In:* Cagan, R.H. and Kare, M.R. eds. *Biochemistry of taste and olfaction.* Academic Press, New York, 7-31.

Healy, T.P. and Jepson, P.C., 1988. The location of floral nectar sources by mosquitoes: the long-range responses of *Anopheles arabiensis* Patton (Diptera: Culicidae) to *Achillea millefolium* flowers and isolated floral odour. *Bulletin of Entomological Research,* 78 (4), 651-657.

Hildebrand, J.G. and Shepherd, G.M., 1997. Mechanisms of olfactory discrimination: converging evidence for common principles across phyla. *Annual Review of Neuroscience,* 20, 595-631.

Hodgkin, J., Plasterk, R.H. and Waterston, R.H., 1995. The nematode *Caenorhabditis elegans* and its genome. *Science,* 270 (5235), 410-414.

Holt, R.A., Subramanian, G.M., Halpern, A., et al. 2002. The genome sequence of the malaria mosquito *Anopheles gambiae. Science,* 298 (5591), 129-149.

Jacquin, J.E. and Merlin, C., 2004. Insect olfactory receptors: contributions of molecular biology to chemical ecology. *Journal of Chemical Ecology,* 30 (12), 2359-2397.

Jurenka, R.A., 1996. Signal transduction in the stimulation of sex pheromone biosynthesis in moths. *Archives of Insect Biochemistry and Physiology,* 33 (3/4), 245-258.

Karlson, P. and Butenandt, A., 1959. Pheromones (ectohormones) in insects. *Annual Review of Entomology,* 4, 39-58.

Kessler, A., Halitschke, R. and Baldwin, I.T., 2004. Silencing the jasmonate cascade: induced plant defenses and insect populations. *Science,* 305 (5684), 665-668.

Lacroix, R., Mukabana, W.R., Gouagna, L.C., et al. 2005. Malaria infection increases attractiveness of humans to mosquitoes. *PLoS Biology,* 3 (9), 1590-1593.

McCall, P.J. and Kelly, D.W., 2002. Learning and memory in disease vectors. *Trends in Parasitology,* 18 (10), 429-433.

Rafaeli, A., 2005. Mechanisms involved in the control of pheromone production in female moths: recent developments. *Entomologia Experimentalis et Applicata,* 115 (1), 7-15.

Reddy, G.V.P. and Guerrero, A., 2004. Interactions of insect pheromones and plant semiochemicals. *Trends in Plant Science,* 9 (5), 253-261.

Restrepo, D., 2004. What the frog s nose tells the frog s brain. *Journal of General Physiology,* 123 (2), 97-98.

Robertson, H.M., Warr, C.G. and Carlson, J.R., 2003. Molecular evolution of the insect chemoreceptor gene superfamily in *Drosophila melanogaster. Proceedings of the National Academy of Sciences of the United States of America,* 100 (Suppl. 2), 14537-14542.

Rudolfs, W., 1922. Chemotropism of mosquitoes. *Bulletin of the New Jersey Agricultural Experimental Station,* 367, 4-23.

Schoonhoven, L.M., Van Loon, J.J.A. and Dicke, M., in press. *Insect-plant biology.* 2nd edn. Oxford University Press, Oxford.

Stoddart, D.M., 1990. *The scented ape: the biology and culture of human odour.* Cambridge University Press, Cambridge.

Stowe, M.K., Turlings, T.C., Loughrin, J.H., et al. 1995. The chemistry of eavesdropping, alarm, and deceit. *Proceedings of the National Academy of Sciences of the United States of America,* 92 (1), 23-28.

Takken, W. and Knols, B.G.J., 1999. Odor-mediated behavior of Afrotropical malaria mosquitoes. *Annual Review of En tomology,* 44, 131-157.

Takken, W., Van Loon, J.J.A. and Adam, W., 2001. Inhibition of host-seeking response and olfactory responsiveness in *Anopheles gambiae* following blood feeding. *Journal of Insect Physiology,* 47 (3), 303-310.

Thaler, J.S., 2002. Effect of jasmonate-induced plant responses on the natural enemies of herbivores. *Journal of Animal Ecology,* 71 (1), 141-150.

Thornsteinson, A.J. and Brust, R.A., 1962. The influence of flower scents on aggregations of caged adult *Aedes aegypti. Mosquito News,* 22 (4), 348-351.

Traniello, J.F.A. and Robson, S.K., 1995. Trail and territorial communication in social insects. *In:* Cardé, R.T. and Bell, W.J. eds. *Chemical ecology of insects. vol. 2*. Chapman & Hall, New York, 241-286.

Tscharntke, T., Klein, A.M., Kruess, A., et al. 2005. Landscape perspectives on agricultural intensification and biodiversity: ecosystem service management. *Ecology Letters,* 8 (8), 857-874.

Turlings, T.C., Loughrin, J.H., McCall, P.J., et al. 1995. How caterpillar-damaged plants protect themselves by attracting parasitic wasps. *Proceedings of the National Academy of Sciences of the United States of America,* 92 (10), 4169-4174.

Van Zandt, P.A. and Agrawal, A.A., 2004. Community-wide impacts of herbivore-induced plant responses in milkweed (*Asclepias syriaca*). *Ecology,* 85 (9), 2616-2629.

Vet, L.E.M., 1996. Parasitoid foraging: the importance of variation in individual behaviour for population dynamics. *In:* Floyd, R.B., Sheppard, A.W. and De Barro, P.J. eds. *Frontiers of population ecology*. CSIRO Publishing, Collingwood, 245-256.

Vet, L.E.M. and Dicke, M., 1992. Ecology of infochemical use by natural enemies in a tritrophic context. *Annual Review of Entomology,* 37 (1), 141-172.

Visser, J.H., 1986. Host odor perception in phytophagous insects. *Annual Review of Entomology,* 31, 121-144.

Vosshall, L.B., 2003. Putting smell on the map. *Trends in Neurosciences,* 26 (4), 169-170.

Wertheim, B., Van Baalen, E.J., Dicke, M., et al. 2005. Pheromone-mediated aggregation in nonsocial arthropods: an evolutionary ecological perspective. *Annual Review of Entomology,* 50, 321-346.

Ylonen, H., Sundell, J., Tiilikainen, R., et al. 2003. Weasels (*Mustela nivalis nivalis*) preference for olfactory cues of the vole (*Clethrionomys glareolus*). *Ecology,* 84 (6), 1447-1452.

CHAPTER 2

CHEMICAL COMMUNICATION

Five major challenges in the post-genomics age

DUSTIN J. PENN

Konrad Lorenz Institute for Ethology, Austrian Academy of Sciences, Savoyenstraße 1a, 1160 Vienna, Austria. E-mail: D.Penn@klivv.oeaw.ac.at

Abstract. Chemical signals play an important role in the behaviour of most, if not all, organisms, but we still have much to learn about this mode of communication. Here I examine some of the major challenges to understanding chemical communication, especially for vertebrates, and consider how genomics, proteomics, metabolomics, and other '-omics' sciences and technologies provide new opportunities to address many of these challenges. First, one of the major challenges of this field is to better understand the kinds of information chemical signals provide. A second challenge is to unravel the proximate mechanisms that control chemical communication (i.e., the production and composition of chemosignals and olfactory recognition). Progress has been advancing rapidly in these areas, especially since the genes that encode odorant receptors were discovered, but there is still much to learn. Third, most research is focused on mechanisms, but there are major unsolved questions regarding the evolution of chemical communication. In particular, we still do not know how signals can evolve to become honest and reliable. A fourth major challenge is to better understand the role of chemical communication in the behaviour of our own species, and integrate this work into the social sciences. The final major challenge is to develop a field of applied chemical signalling that addresses problems in agriculture, medicine and the environment. In particular, we need to determine how chemical pollutants in our environment disrupt biological chemical signalling systems and potentially affect the health of humans and wildlife (ethotoxicology and ecotoxicology).
Keywords: pheromones; ecogenomics; ethogenomics; sociogenomics; endocrine disruptor chemicals

INTRODUCTION

Chemical communication is a universal feature of life that occurs at all levels of biological organization, including regulation of cells and organs within the body, as well as social behaviour and ecological interactions among individuals (Agosta 1992). The terminology used for communication is constantly evolving, and so for clarification, I will use the term semiochemicals for chemicals used for information conveyance, and the term pheromones for those semiochemicals used for intraspecific communication. Pheromones play an important role in the behaviour of a wide variety of organisms, from moths to elephants (Wyatt 2003). Chemical cues provide several possible advantages compared to other sensory modalities (Doty 1986). They can be used in situations in which visual cues are unavailable, for

M. Dicke and W. Takken (eds.), Chemical Ecology: From Gene to Ecosystem, 9-18.

example, and they provide spatial information, such as space occupancy. A problem with chemical signals is that they are more difficult to observe or measure than visual or acoustic ones, and therefore they remain less understood. There are many unsolved mysteries about chemical communication. My aim here is to review some of the main challenges for chemical-communication research, with an emphasis on mammals and other vertebrates, and consider how genomics and other '-omics' technologies offer opportunities to solve some of these problems.

DETERMINING THE KINDS OF INFORMATION ENCODED IN CHEMICAL SIGNALS

Odour can reveal much information about an individual, including sex, diet, social status, individual and group identity, reproductive condition, age, health, fear and other emotional states (Wyatt 2003). Scent marks and many other semiochemicals can be thought of as *extended phenotypes* (Dawkins 1983), though we know little about the genetics of semiochemical production. It has been suggested that pheromones and other chemical cues provide indicators that advertise a male's health and resistance to disease to potential mates, functionally analogous to the colourful secondary sexual traits of birds (Penn and Potts 1998). An individual's scent not only provides an indicator of infection, it also appears to indicate the activation of immunity (Zala et al. 2004). It is unclear how this occurs, though odour has long been used to diagnose a variety of diseases (Penn and Potts 1998). Odour also provides an indicator for assessing genetic relatedness and genetic compatibility of potential mates (Penn 2002), though it is also unclear how this occurs. Wilson (1970) suggested that in vertebrates individual identification is the most important message used in chemical communication, and there has been an increasing interest in determining whether individuals have unique chemical fingerprints or odourtypes (Beauchamp and Yamazaki 2003).

UNRAVELLING THE MECHANISMS CONTROLLING CHEMICAL COMMUNICATION

Chemical signals convey an amazing amount of information, and so one of the major challenges is to determine how this occurs. In particular, we need to know more about the compounds that are involved, how they are produced, and how olfactory organs are able to 'decode' information from chemical signals.

Determining the chemical composition of semiochemicals

The vast majority of semiochemicals of interest remain to be chemically identified. A variety of techniques are used for chemical analyses, especially *combined gas chromatography and mass spectroscopy* (GC-MS). Finding biologically active compounds, however, is like finding needles in haystacks, and one way to narrow down the possibilities is to use the olfactory organs of animals as sensors for determining the bioactivity of compounds. In arthropods the bulk of olfactory neurons are contained in filamentous antennae from which so-called 'electroantennograms' (EAGs) can conveniently be recorded. This method has successfully been used for the identification of pheromones in numerous insects. More recently electrophysiological activity recorded from the olfactory bulb has been used as a biosensor signal in mammals (Lin et al. 2005). Chemical identification of active components among the many peaks in a complex chromatogram is still not an easy task, however, and conducting library searches to get clues about the identity of a compound from its mass spectrum (and retention time) is just one step in this process. After chemical identification, pheromones can then be synthesized using techniques in organic chemistry, and then confirmed using behavioural or neurological bioassays. Another problem is that chemical analyses are often just qualitative, identifying the presence or absence of compounds, even though there is potentially a great deal of information contained in quantitative levels of odorants, the ratios of multiple components (multicomponent pheromones and multivariate fingerprints), and the dynamic expression of these compounds. Fortunately, though, new developments in analytical chemistry are making it possible to obtain quantitative chromatographic data. This is largely due to improvements in solventless sampling techniques, such as open-tubular trapping (OTT), solid-phase microextraction (SPME) and particularly stir-bar-sorptive extraction (SBSE) (Baltussen et al. 2002; 1999; Soini et al. 2005). Moreover, recent advances in *chemometrics* offer powerful statistical analyses, such as pattern recognition, that are used to discover and quantify compounds of interest in complex chromatographic profiles (Brereton 2003). Chemical identification of the compounds is necessary; but it is not sufficient as we also need to understand how these compounds are produced.

Determining how semiochemicals are produced

This problem is becoming easier to solve due to the increasing availability of high-throughput tools from genomics, proteomics, metabolomics and other -omics technologies (Box 1). These have proved useful for determining the structure of carrier molecules (lipocalins) that bind and transport volatile compounds to urine and saliva (Timm et al. 2001; Spinelli et al. 2002). Determining the metabolic origins of individual odours is likely to be complicated because complex communities of commensal microflora probably play an important role (Albone et al. 1977). Commensal microflora is still not well described for any species, largely because the majority cannot be cultured in the laboratory. However, recently developed molecular genetic tools are successfully being applied to solve this problem (PCR-DGGE profiling) (Tannock 2002). Identification of compounds and determining the origin of their production will help to understand the underlying mechanisms; however, we also need to better understand the receiver side of communication.

Determining the molecular basis for olfaction

Determining how chemical signals are detected, processed, and how they trigger the perception of smell has been an extremely difficult problem. Olfaction has long been the least understood of all the senses, but progress has been advancing rapidly, especially since Buck and Axel discovered the olfactory receptor (OR) genes (a discovery for which they recently received a Nobel Prize) (Buck and Axel 1991). Tools from genomics and other -omics sciences have subsequently been helping to improve our understanding of olfaction (Young and Trask 2002). OR proteins bind odorant molecules and then initiate neural responses that trigger the perception of smell (De Bruyne in press). OR genes comprise one of the largest known gene families, with 900 (humans) to 1500 (mice) loci, scattered throughout the genome. The current paradigm is that each olfactory neuron expresses a single allele of a single OR gene through some sort of allelic selection process during development, but an exception has recently been reported for *Drosophila* (Goldman et al. 2005). It is unclear how the nervous system turns signals from olfactory neurons into the perception of smell – and how it integrates input from multiple sensory modalities, though these problems are gradually being solved. Unravelling the molecular basis for olfaction will be a major advancement (De Bruyne in press), but even this is not sufficient to understand chemical communication fully because we ultimately need to explain how such complex mechanisms evolved. For example, OR genes are highly polymorphic in sequences and copy numbers, and yet it is completely unclear how natural selection maintains this enormous diversity.

DETERMINING HOW CHEMICAL SIGNALS EVOLVE AND CONVEY RELIABLE INFORMATION

One of the central problems for the study of animal communication is explaining why signals can evolve to become honest and reliable (Maynard-Smith and Harper 2003). Not all signals are honest, of course, as there are many examples of deceit and manipulation. For example, male moths are attracted to the pheromones of conspecific females, and bolas spiders in two independent lineages have evolved the ability to synthesize moth pheromones, which they use to lure male moths (Stowe et al. 1995). However, signals are usually reliable because otherwise receivers would ignore them, and the signalling system would cease to exist. Signalling should lead to a dynamic co-evolutionary 'arms race' between signallers and receivers, with signallers evolving ways to cheat and manipulate others, and receivers evolving mechanisms to resist manipulation and 'mind-read' signallers (Krebs and Dawkins 1984). This arms race model is surely correct, at least when signallers and receivers do not have mutual interests, though it has not been tested to my knowledge.

There are at least three explanations for the evolution of stable and reliable signalling. First, *the handicap principle* suggests the counter-intuitive notion that honest signals can evolve precisely *because* they are costly to produce and cannot be faked (Zahavi and Zahavi 1997). It has been suggested that chemical signals are strategic handicaps that provide honest indicators of a male's quality to rivals and potential mates (Penn and Potts 1998). Contrary to what has become widely assumed, however, the handicap principle is not the only explanation for reliability. A second explanation for how reliable signals can evolve, is when the signal and receiver have *common interests* in the outcome of their interaction. For example, species recognition signals used in mate choice to avoid hybridization can be honest and cheap because there is no benefit to cheating. Similarly, signals among cells within the body need not be costly to be honest as they generally have shared interests. Third, signals can be honest when they provide an *index* of some aspect of the organism, such as size, that is unmodifiable and therefore the signaller simply cannot lie. For example, it has been suggested that odour cues provide an honest indicator of health and disease because the volatile metabolic by-products of an immune response and disease are impossible to disguise (Penn and Potts 1998). The task of determining the reliability and costs of chemical signalling has only just begun, which includes measuring the energetic costs, and ecological costs, such as exposing the owner to greater risks of predation or parasitism.

DETERMINING THE ROLE OF CHEMICAL COMMUNICATION IN HUMAN BEHAVIOUR

Although the existence of human pheromones remains controversial, there is increasing evidence that volatile chemical signals influence human behaviour (Hays 2003; Stoddart 1990; Wysocki and Preti 2004). We need to know more about the types of information that humans convey by scent, and especially how other individuals respond to chemical signals. For example, a woman's scent indicates whether she is ovulating or not (Singh and Bronstad 2001), though we do not know how this affects the behaviour of other female of male individuals. An individual's odour changes when a fearful situation is perceived (Ackerl et al. 2002), but it is not known whether this triggers fear or anxiety in other individuals (such as 'fear pheromones'). There are very few examples of chemical signals affecting another individual's physiology or behaviour. The best examples are some unknown pheromones that somehow synchronize women's menstrual cycles (McClintock 1971), and yet it is unclear whether menstrual synchrony is functional or even occurs under natural situations. There appear to be pheromones that induce hormonal changes and trigger changes in emotions and moods (Jacob and McClintock 2000). There is evidence that odour plays a role in kin recognition, including a study that found that infants move towards scent from their mother's breast (Porter 1998), individual recognition, and mate choice (Penn 2002).

Although they appear to exist, no human pheromones have been chemically identified to date, and so this presents an important challenge. This challenge is similar to identifying the active ingredients in useful herbal medicines, such as isolating digitoxin in foxglove (Wysocki and Preti 2004). The human axillae are probably functional analogues to scent glands of other mammals (Stoddart 1990). For example, in humans protein (lipocalin) molecules carry odorants to the axillae, where they are metabolized and made volatile by commensal microflora (Spielman et al. 1995), which seems to be analogous to major urinary proteins (MUPs) and other carrier proteins used by other mammals. The greatest progress has been unravelling the mechanisms controlling olfaction, and there is now overwhelming evidence that the vomeronasal organ (VNO) is not functional in human adults (Wysocki and Preti 2004). This means that human pheromones must be detected by the main olfactory bulb, despite popular misconceptions that pheromones are only detected by the VNO.

Chemical communication in humans has largely been ignored, though this situation is changing and human pheromones are attracting increasing attention. Integrating chemical-communication research into the social sciences will be easier as the artificial barriers between the human and natural sciences are breaking down. There will likely be more interest in chemical communication as researchers find more applications for our own species, such as in medicine.

DETERMINING HOW POLLUTANTS DISRUPT CHEMICAL SIGNALS

One of the most difficult challenges is to use our understanding of chemical communication to address applied problems, such as in medicine, agriculture and the environment. For example, in medicine, artificial chemical sensors or e-noses are currently being developed to diagnose diseases, such as cancer, via a patient's breath or urinary odour (Turner and Magan 2004).

Chemical-communication research has surprising implications for toxicology. There are an increasing number of chemical pollutants in our environment and in our bodies, and many of these are not toxic or carcinogenic, and yet they cause numerous other problems, such as altering sexual development. The problem is that they are chemically similar to the body's own hormones (estrogen mimics) or they otherwise disrupt the body's own internal chemical signals (Colborn et al. 1996b). These so-called endocrine-disrupting chemicals (or EDCs) impact endocrine, neural, immune and behavioural responses. In an outstanding book on the topic, called *Our Stolen Future,* Colborn et al. (1996b) point out that "The key concept in thinking about this kind of toxic assault is *chemical messages*. Not poisons, not carcinogens, but *chemical messages*." (p. 204; italics added). Since then, several studies have found that pheromones and other semiochemicals are negatively affected by EDCs (Zala and Penn 2004; Fox 2004). The impact of these endocrine-disrupting chemicals for humans and wildlife is still controversial, though this has become the focus of the new interdisciplinary fields of ecotoxicology and ethotoxicology. Recently, researchers have increasingly been applying tools from -omics technologies to address problems in ecotoxicology (Robertson 2005).

CONCLUSIONS

Many vertebrate species, including our own, use chemistry to communicate, though exactly how is still rather mysterious. The increasing number of new tools available in analytical chemistry, chemometrics, molecular biology, and genetics, are leading to exciting new discoveries. These new technologies provide unprecedented opportunities, but they also create a new set of problems. For instance, we need to find ways to analyse statistically the enormous amount of complex data generated from chromatographic profiles and DNA microarrays. Also, they will not replace the crucial role of theory: as one researcher, Christer Löfstedt, points out, "to obtain an interesting answer from your research, it helps to ask an interesting question!". There are numerous other important problems in chemical communication that I did not address here. Perhaps, the most important problem is clarifying all of the links that make up chemical communication, from pheromone production by the emitter on one end, to olfactory reception by receivers on the other, in a single model organism, such as house mice (Emes et al. 2004). A more integrated understanding of chemical communication will require insights into ecology and evolution. The problem is that we still know little about the ecology and evolution of house mice and other model organisms, as the importance of ecology and evolution for understanding the 'design' of these organisms and their genomes is not generally appreciated. Therefore, organisms whose ecology and evolution are well-studied would make excellent subjects for a genome project, and could become models for studying chemical communication.

Box 1. *Chemical communication in the post-genomic era*

The increasing availability of high-throughput tools from genomics and other -omics sciences and technologies allows researchers to measure gene expression (transcriptomics) and to determine protein structure (proteomics) and metabolic profiles (metabolomics). These tools help to identify gene products (transcripts, proteins, metabolites) in a sample, and examine quantitative dynamics in biological systems (Kell 2004).

Genomics is already being applied to address ecological questions about chemical communication (ecogenomics) (Berenbaum and Robinson 2003; Dicke et al. 2004). These -omics technologies are just beginning to be applied to address animal behaviour (Pennisi 2005), the evolution of behaviour (behavioural ecology) (Feder and Mitchell-Olds 2003; Fitzpatrick et al. 2005), and the evolution of social behaviour (sociogenomics) (Robertson 2005). Sociogenomics is a sub-discipline of behavioural genomics, or what could be called 'ethological genomics' or 'ethogenomics'. Combined with improved phenotyping tools, ethogenomics and sociogenomics have the potential to become core disciplines for chemical-communication research, linking chemistry and physiology on one end with ecology and evolution on the other.

The various -omics sciences and technologies offer new opportunities to investigate chemical communication; however, they also generate such massive datasets that new methods for managing, processing and analysing data are required (bioinformatics). Sir Peter Medawar (1982) argued that "...there is an epoch in the growth of a science during which facts accumulate faster than theories can accommodate them..." (p. 29). The post-genomics age appears to be just such an epoch, as it is becoming increasingly difficult to keep up with the explosion of data and facts! Still, to better understand highly complex systems, such as the genome and metabolism, proper data handling and analysis are crucial, and there is increasing interest in applying modelling techniques from *systems biology* (Kell 2004; Provart and McCourt 2004). Perhaps theoretical approaches from systems biology could also help to understand more complex problems in chemical communication.

ACKNOWLEDGEMENTS

I wish to thank Marcel Dicke and Willem Takken, and two anonymous reviewers for their useful comments and corrections. I also wish to thank my post-doctoral student, Kerstin Musolf for her input.

REFERENCES

Ackerl, K., Atzmüller, A. and Grammer, K., 2002. The scent of fear. *Neuro Endocrinology Letters,* 23 (2), 79-84.

Agosta, W.C., 1992. *Chemical communication: the language of pheromones.* Scientific American Library, New York.

Albone, E.S., Gosden, P.E. and Ware, G.C., 1977. Bacteria as a source of chemical signals in mammals. *In:* Muller-Schwarze, D. and Mozell, M. eds. *Chemical signals in vertebrates.* Plenum Press, New York, 35-44.

Baltussen, E., Cramers, C.A. and Sandra, P.J.F., 2002. Sorptive sample preparation: a review. *Analytical and Bioanalytical Chemistry,* 373 (1/2), 3-22.

Baltussen, E., Sandra, P., David, F., et al. 1999. Stir bar sorptive extraction (SBSE), a novel extraction technique for aqueous samples: theory and principles. *Journal of Microcolumn Separations,* 11 (10), 737-747.

Beauchamp, G.K. and Yamazaki, K., 2003. Chemical signalling in mice. *Biochemical Society Transactions,* 31 (1), 147-151.

Berenbaum, M.R. and Robinson, G.E., 2003. Chemical communication in a post-genomic world. *Proceedings of the National Academy of Sciences of the United States of America,* 100, 14513-14513.

Brereton, R.G., 2003. *Chemometrics: data analysis for the laboratory and chemical plant.* Wiley, Chichester.

Buck, L. and Axel, R., 1991. A novel multigene family may encode odorant receptors: a molecular basis for odor recognition. *Cell,* 65 (1), 175-187.

Colborn, T., Dumanoski, D. and Myers, J.P., 1996a. *Our Stolen Future: Are We Threatening Our Fertility, Intelligence, and Survival? A Scientific Detective Story.* Dutton, New York.

Colborn, T., Dumanoski, D. and Myers, J.P., 1996b. *Our stolen future: are we threatening our fertility, intelligence, and survival? A scientific detective story.* Little, Brown and Company, Boston.

Dawkins, R., 1983. *The extended phenotype: the gene as the unit of selection.* Oxford University Press, Oxford.

De Bruyne, M., in press. Visualizing a fly's nose: genetic and physiological techniques for studying odour coding in *Drosophila. In:* Dicke, M. and Takken, W. eds. *Chemical ecology: from gene to ecosystem.* Springer, Dordrecht. Wageningen UR Frontis Series No. 16. [http://library.wur.nl/frontis/chemical_ecology/08_de_bruyne.pdf].

Dicke, M., Van Loon, J.J.A. and De Jong, P.W., 2004. Ecogenomics benefits community ecology. *Science,* 305 (5684), 618-619.

Doty, R.L., 1986. Odor-guided behavior in mammals. *Experientia,* 42 (3), 257-271.

Emes, R.D., Beatson, S.A., Ponting, C.P., et al. 2004. Evolution and comparative genomics of odorantand pheromone-associated genes in rodents. *Genome Research,* 14 (4), 591-602.

Feder, M.E. and Mitchell-Olds, T., 2003. Evolutionary and ecological functional genomics. *Nature Reviews Genetics,* 4 (8), 651-657.

Fitzpatrick, M.J., Ben-Shahar, Y., Smid, H.M., et al. 2005. Candidate genes for behavioural ecology. *Trends in Ecology and Evolution,* 20 (2), 96-104.

Fox, J.E., 2004. Chemical communication threatened by endocrine-disrupting chemicals. *Environmental Health Perspectives,* 112 (6), 648-653.

Goldman, A.L., Van der Goes-Van Naters, W., Lessing, D., et al. 2005. Coexpression of two functional odor receptors in one neuron. *Neuron,* 45 (5), 661-666.

Hays, W.S.T., 2003. Human pheromones: have they been demonstrated? *Behavioral Ecology and Sociobiology,* 54 (2), 89-97.

Jacob, S. and McClintock, M.K., 2000. Psychological state and mood effects of steroidal chemosignals in women and men. *Hormones and Behavior,* 37 (1), 57-78.

Kell, D.B., 2004. Metabolomics and systems biology: making sense of the soup. *Current Opinion in Microbiology,* 7 (3), 296-307.

Krebs, J.R. and Dawkins, M.S., 1984. Animal signals: mind-reading and manipulation. *In:* Krebs, J.R. and Davies, N.B. eds. *Behavioural ecology: an evolutionary approach.* 2nd edn. Blackwell, Oxford, 380-402.

Lin, D.Y., Zhang, S.Z., Block, E., et al. 2005. Encoding social signals in the mouse main olfactory bulb. *Nature,* 434 (7032), 470-477.

Maynard-Smith, J. and Harper, D., 2003. *Animal signals.* Oxford University Press, Oxford.

McClintock, M.K., 1971. Menstrual synchrony and suppression. *Nature,* 229 (5285), 244-245.

Medawar, P., 1982. *Pluto's republic: incorporating the art of the soluble and induction and intutition.* Oxford University Press, Oxford.

Penn, D. and Potts, W.K., 1998. Chemical signals and parasite-mediated sexual selection. *Trends in Ecology and Evolution,* 13 (10), 391-396.

Penn, D.J., 2002. The scent of genetic compatibility: sexual selection and the major histocompatibility complex. *Ethology,* 108 (1), 1-21.

Pennisi, E., 2005. Genetics: a genomic view of animal behavior. *Science,* 307 (5706), 30-32.

Porter, R.H., 1998. Olfaction and human kin recognition. *Genetica,* 104 (3), 259-63.

Provart, N.J. and McCourt, P., 2004. Systems approaches to understanding cell signaling and gene regulation: commentary. *Current Opinion in Plant Biology,* 7 (5), 605-609.

Robertson, D.G., 2005. Metabonomics in toxicology: a review. *Toxicological Sciences,* 85 (2), 809-822.

Singh, D. and Bronstad, P.M., 2001. Female body odour is a potential cue to ovulation. *Proceedings of the Royal Society of London. Series B. Biological Sciences,* 268 (1469), 797-801.

Soini, H.A., Bruce, K.E., Wiesler, D., et al. 2005. Stir bar sorptive extraction: a new quantitative and comprehensive sampling technique for determination of chemical signal profiles from biological media. *Journal of Chemical Ecology,* 31 (2), 377-392.

Spielman, A.I., Zeng, X.N., Leyden, J.J., et al. 1995. Proteinaceous precursors of human axillary odor: isolation of two novel odor-binding proteins. *Experientia,* 51 (1), 40-47.

Spinelli, S., Vincent, F., Pelosi, P., et al. 2002. Boar salivary lipocalin: three-dimensional X-ray structure and androstenol/androstenone docking simulations. *European Journal of Biochemistry,* 269 (10), 2449-2456.

Stoddart, D.M., 1990. *The scented ape: the biology and culture of human odour.* Cambridge University Press, Cambridge.

Stowe, M.K., Turlings, T.C., Loughrin, J.H., et al. 1995. The chemistry of eavesdropping, alarm, and deceit. *Proceedings of the National Academy of Sciences of the United States of America,* 92 (1), 23-28.

Tannock, G.W., 2002. Analysis of the intestinal microflora using molecular methods. *European Journal of Clinical Nutrition,* 56, S44-S49.

Timm, D.E., Baker, L.J., Mueller, H., et al. 2001. Structural basis of pheromone binding to mouse major urinary protein (MUP-I). *Protein Science,* 10 (5), 997-1004.

Turner, A.P. and Magan, N., 2004. Electronic noses and disease diagnostics. *Nature Reviews Microbiology,* 2 (2), 161-166.

Wilson, E.O., 1970. Chemical communication within animal species. *In:* Sondheimer, E. and Simeone, J.B. eds. *Chemical Ecology.* Academic Press, New York, 133-155.

Wyatt, T.D., 2003. *Pheromones and animal behaviour: communication by smell and taste.* Cambridge University Press, Cambridge.

Wysocki, C.J. and Preti, G., 2004. Facts, fallacies, fears, and frustrations with human pheromones. *Anatomical Record.Part A. Discoveries in Molecular Cellular and Evolutionary Biology,* 281A (1), 1201-1211.

Young, J.M. and Trask, B.J., 2002. The sense of smell: genomics of vertebrate odorant receptors. *Human Molecular Genetics,* 11 (10), 1153-1160.

Zahavi, A. and Zahavi, A., 1997. *The handicap principle: a missing piece of Darwin's puzzle.* Oxford University Press, Oxford.

Zala, S.M. and Penn, D.J., 2004. Abnormal behaviours induced by chemical pollution: a review of the evidence and new challenges. *Animal Behaviour,* 68, 649-664.
Zala, S.M., Potts, W.K. and Penn, D.J., 2004. Scent marking in mice provides honest signals of health and infection. *Behavioral Ecology,* 15 (2), 338-344.

CHAPTER 3

PLANT–INSECT INTERACTIONS IN THE ERA OF CONSOLIDATION IN BIOLOGICAL SCIENCES

Nicotiana attenuata as an ecological expression system

ANDRÉ KESSLER

Department of Ecology and Evolutionary Biology, Cornell University, Ithaca, NY 14853, USA. E-mail: ak357@cornell.edu

Abstract. The past decades have seen an intense development of organismal biology and genomics of individual species on the one hand, and population biology and evolutionary ecology on the other. While the great discoveries fuelled by the current model systems will continue over the next decades, more and more discoveries will occur at the interface between different biological disciplines. It is through such integrative approaches that the mechanisms of evolution and adaptation will be revealed. The study of plant–insect interactions, exemplary among such integrative research fields, unifies research efforts on the cellular and organismal level with those on the whole-plant and community level. Recent studies on the wild tobacco plant *Nicotiana attenuata* illustrate both the value of using genetic and molecular tools in ecological research and the importance of profound natural-history knowledge when studying plant–insect interactions.

Keywords: induced plant responses; herbivory; plant defence; jasmonate signalling

INTRODUCTION – THE MODERN CONSOLIDATION IN BIOLOGY

We are in the midst of what is widely regarded as the century of biology. Life science is already influencing multiple aspects of the modern economy and is expected to move to the forefront of all the sciences. Biotechnology, projected to become an unparalleled industrial mainstay, already touches everyone's daily life. Its increasing importance even prompts university departments in the traditional engineering disciplines to offer life science as part of their curricula (Friedman 2001). The notion of the dominant role of life sciences in modern research and the economy is to a great extent based on the breathtakingly fast advances in genetics and molecular biology. It is projected that through this progress we will eventually be enabled to reveal how cells, organisms and ecosystems function. But will we?

The disproportionately high allocation of workforce and financial resources to genetics and molecular biology has led to an apparent under-representation of subjects such as natural history and organismal biodiversity in our biological curricula. Greene (2005) argues that scientific theories help us to study nature better

M. Dicke and W. Takken (eds.), Chemical Ecology: From Gene to Ecosystem, 19-37.

through summarizing current knowledge and formulating hypotheses. Nonetheless, such theories cannot themselves replace discoveries of new organisms or new facts about organisms. The continuous study of species' natural history can help to reset research cycles and may change the hypothesis testing that underlies conceptual progress in science. Following this integrative notion, a profound understanding of ecological and evolutionary processes can be found only at the interface between different biological research domains. Scientific progress will ultimately be based on unification rather than fragmentation of knowledge (Kafatos and Eisner 2004).

On the threshold of the biological era, the life sciences are at the inception of a profound transformation by starting a process of consolidation. The life sciences have long formed two major domains, one reaching from the molecule to the organism, the other bringing together population biology, biodiversity study and ecology. Kafatos and Eisner (2004) argue that these domains, kept separate, no matter how fruitful, cannot deliver on the full promise of modern biology. Only the unification of the two research domains can lead to a full appreciation of life's complexity from the molecule to the biosphere or, indeed, maximize the benefits of biological research for medicine, industry, agriculture or conservation biology. Many researchers and academic institutions have recognized the necessity for unification and created research environments with integrative collaborations of researchers representing different disciplines and teaching programmes that emphasize multi-disciplinary approaches.

Chemical ecology is a discipline that emerged during the past half century and is by definition an integrative research field. It is driven by the recognition that organisms of diverse kinds make use of chemical signals to interact (Karban and Baldwin 1997). The original endeavour to decipher the chemical structure and the information content of the mediating molecules as well as the ecological consequences of signal transduction is now receiving a major directional addition, the modern domain of molecular biology (Eisner and Berenbaum 2002). It promises an understanding of the molecular and genetic mechanisms of biological signal transduction in species interactions, which can help to ultimately understand the evolution of complex species interactions.

The study of plant–insect interactions is an excellent example of the success of the modern approaches taken in chemical and molecular biology (e.g.,Walling 2000; Berenbaum 2002; Kessler and Baldwin 2002; Dicke and Hilker 2003; Hartmann 2004). It is a fast-growing field within the research of organismal interactions to a great extent because the results can readily be applied in modern agriculture and therefore have a potentially high economic value (Khan et al. 2000). In the following sections I will summarize selected characteristics of induced plant responses to herbivory that at the same time define integrative focus directions in this research field, ranging from the physiological and ecological costs and consequences to the cellular signalling crosstalks that result in the elicitation of plant defences. In addition I will summarize studies of the complex multitrophic interactions of the wild tobacco plant *Nicotiana attenuata* (Torr. ex Watts) with its insect community that emphasize the potential role of induced plant defences in structuring arthropod communities and the value of using molecular and chemical analytical tools in ecological research.

PLANT–INSECT INTERACTIONS

Chemical communication can be studied at various levels of integration reaching from the expression of genes involved in biosynthesis of signal molecules to ecological consequences of the resulting organismal interactions on the community level. When studying plant–insect interactions we observe an exchange of signals that reciprocally influence the interacting partners and consequently include a complex crosstalk across all the levels of integration. Moreover, plant–insect interactions are played out in an arena that is much bigger than the plant itself. It includes interferences on the cellular level that have been extensively studied in plant–pathogen interactions (e.g., Lam et al. 2001; Van Breusegem et al. 2001) as well as interactions at the whole-plant and the community level. The latter result from multitrophic and inter-guild interactions, which are frequently mediated by the plants' chemical defences (Agrawal 2000; Dicke and Van Loon 2000; Karban and Agrawal 2002; Kessler and Baldwin 2002).

The fitness costs of plant defences

Plants have myriad ways to defend themselves against their attackers, including the production of defensive chemicals such as secondary metabolites and defensive proteins (Duffey and Stout 1996). The evolutionary arms race between plants and herbivorous insects has early on been suggested as one of the driving forces of the chemical diversity in the plant kingdom. Ehrlich and Raven (1964) coined the term 'coevolution' and stimulated entomological studies of how plants and insects influence each other's evolutionary trajectories. In the notion of their coevolutionary hypothesis a plant species' evolutionary innovation of new defensive compounds results in the exclusion of potential herbivores, which in turn will be strongly selected to tolerate or detoxify the new plant compounds. The counter-defences of insects are by no measure less diverse than the plant defences, and reach from the elicitation of changes in plant morphology (Sopow et al. 2003) to the sequestration of plant secondary metabolites and their use for the insects' own defences against natural enemies (Hartmann 2004).

The production of plant defence traits when they are not needed (e.g., in absence of herbivores) incurs significant fitness cost for a number of reasons (Agrawal et al. 1999; Heil and Baldwin 2002). First, the production of secondary metabolites can be costly if fitness-limiting resources are invested (Baldwin 2001; Heil and Baldwin 2002). For example, recent studies on nutrient-rich clay habitats and nutrient-poor white-sand habitats in the Peruvian Amazon region show that immature trees in nutrient-poor habitats are not able to compensate for severe herbivore damage. The nutrient-poor habitat therefore selects for plant species that invest more in defensive secondary-metabolite production at the cost of slower growth (Fine et al. 2004). However, resistance costs can also arise from higher-level ecological processes. For example, specialized herbivores may sequester defensive plant metabolites and use them for their own defence against predators (Karban and Agrawal 2002; Reddy and Guerrero 2004; Hartmann 2004), or compounds that provide defence against generalist herbivores may attract specialist herbivores, which use them as host location signals (Turlings and Benrey 1998). In addition plant defences may disrupt important mutualistic interactions with other insects such as pollinators (Adler et al. 2001) and parasitoids (Campbell and Duffey 1981; Barbosa et al. 1991), and may differently affect the performance of interacting organisms across several trophic levels (Orr and Boethel 1986; Harvey et al. 2003).

Induced plant defences

Constitutively high production of costly defences could only be beneficial for a plant if herbivore pressure is a predictable environmental factor. Unpredictable environments would select for plants that are able to produce a defence only when needed, in the presence of herbivores. Such phenotypically plastic plant responses are referred to as induced defences (Karban and Baldwin 1997). The fitness costs of the production of defensive compounds probably provide the selection pressure behind the evolution of inducible defences. Herbivore-induced plant defences have received a considerable attention in the past few decades, in part because the ecological implications for the plant and its arthropod community are different from those that derive from purely constitutive defences. Induced defences extend plant–insect interactions from the cell and whole-plant level to the community level.

Plants can respond to herbivore damage with the increased production of secondary metabolites or defensive proteins that are categorized by their mode of action (Duffey and Stout 1996). Compounds such as alkaloids, glucosinolates (in combination with myrosinase) and terpenoids function as toxins while proteinase inhibitors and polyphenol oxidases function as anti-digestive or anti-nutritive compounds, respectively. A plant inducing such defences in response to herbivory has a lower nutritive value for subsequently arriving herbivores and therefore reduces the probability of secondary attacks. The plant's metabolic changes may thereby not only affect insects of the same species but may result in cross-resistance effects that affect the herbivore-community composition of this plant (Agrawal 1998; Kessler and Baldwin 2004).

In addition to direct defensive secondary metabolites, plants produce volatile organic compounds (VOCs) in response to herbivore damage. These can function as signals for organisms able to receive and respond to changed odour bouquets. The most studied function of herbivore-induced VOC emission is the attraction of natural enemies such as parasitoids and/or predators to the damaged plant, a process referred to as indirect plant defence (Dicke and Van Loon 2000; Turlings and Benrey 1998). The VOC signal increases the natural enemy's foraging success and therefore facilitates top-down control of the herbivore population. The VOC response can be highly specific. For example, parasitoid wasps can use the specificity of the signal to locate particular hosts or even particular instars of their hosts (Turlings and Benrey 1998). On the other hand, generalist herbivores can also be attracted by single compounds of the VOC bouquet, which are commonly emitted after attack from a diverse set of herbivore species (Kessler and Baldwin 2001). In addition to attracting natural enemies, VOCs can function as direct defences by repelling ovipositing herbivores (De Moraes et al. 2001; Kessler and Baldwin 2001;

2004) or they may be involved in plant–plant interactions (Arimura et al. 2000; Karban et al. 2000).

Indirect plant defences may be compromised by direct plant defences if herbivores are able to sequester secondary plant metabolites and use them for their own defence. A number of studies have shown negative effects of plant secondary metabolites on the third (Campbell and Duffey 1981; Barbosa et al. 1991) and the fourth trophic level (Orr and Boethel 1986; Harvey et al. 2003), and suggest trait-offs between direct and indirect defences. However, direct and indirect defences have rarely been manipulated or characterized in the same experiment. Moreover, the parasitoid performance was only investigated in non-choice experiments. Since herbivore-induced VOC emission is a signal that is very specifically associated with host/prey (Turlings and Benrey 1998) it may provide information not only about the spatial distribution of potential hosts/prey species but also about their quality. Parasitoids or predators of the third trophic level may well be able to differentiate between good and bad hosts and may, in nature, actively avoid hosts which sequester plant metabolites that the natural enemies can not detoxify. Therefore we may more commonly observe a synergism rather than a trade-off between direct and indirect defences in nature because the plant's direct defences may amplify the effects of parasitoid/predator attraction (Kessler and Baldwin 2004, see example below). There is an urgent need to approach this question for the apparent trade-off between direct and indirect defences in native systems without artificial human selection, because answering it provides one of the most important building blocks for utilizing plant defences in sustainable agriculture.

Defensive function of plant secondary metabolites

The biosynthetic pathways involved in the production of secondary metabolites have been or are currently elucidated with impressive speed, and the progress in identification of the underlying genetic and transcriptional mechanisms will only enhance this exploratory process. However, the knowledge about the ecological consequences of induced direct and indirect defences is sketchy and we are far from appreciating the complexity of the arena of plant–insect interactions to its full extent. The defensive function as well as direct physiological or indirect ecological costs of secondary-metabolite production can only be evaluated when the defensive traits can be experimentally manipulated and tested in comparative experiments, ideally in the plants' natural habitats. This can be accomplished by both using chemical elicitors to induce specific plant responses and using mutants or transgenic plants that are not able to produce or over-express a particular defence (Thomas and Klaper 2004). This latter approach is largely restricted to a few model plant species, such as *Arabidopsis thaliana* (e.g., Van Poecke and Dicke 2004; D'Auria and Gershenzon 2005) or a limited number of agricultural crops (e.g., tomato, maize, rice). Thereby the current widespread exposure of genetically modified crop plants that express new defensive compounds, such as *Bacillus turingiensis*-toxin (Bt-toxin), to the natural arthropod community could be used to elucidate principal patterns in the plant–insect coevolutionary process (e.g., the evolution of the insect's resistance to plant toxins) (Tabashnik et al. 1998). However, the conclusions derived from studies in agro-ecosystems may be limited because frequently neither the crop plant nor their herbivores are studied in their native habitats where coevolutionary processes occur or occurred. Similarly limiting is the study of single native species. The induction mechanisms of plant defences may differ among species specifically depending on internal factors, such as signal perception and transduction (elicitation), and external factors, such as the frequently complex web of interacting species on multiple trophic levels and abiotic factors. Thus, the inclusion of additional, preferentially native study systems to survey the diversity of internal and external factors influencing plant–insect interactions, will eventually reveal the general underlying mechanisms, which would allow a sustainable utilization of plant defences in agriculture.

Elicitation of plant responses

Any compound that comes from herbivores and interacts with the plant on a cellular level is a potential elicitor. A series of herbivore-derived elicitors have been isolated from the oral secretion of lepidopteran caterpillars and the oviposition fluid of weevil beetles. The elicitors represent three classes of compounds; lytic enzymes (Mattiacci et al. 1995; Felton and Eichenseer 1999), fatty-acid–amino-acid conjugates (FACs) (Halitschke et al. 2001; Alborn et al. 1997; Pohnert et al. 1999) from caterpillar regurgitant, and bruchins from the oviposition fluid of *Callosobruchus maculatus* (Doss et al. 2000).

Both herbivore feeding and mechanical damage induce plant responses that are systemically propagated throughout the plant or remain locally restricted to the wound site. As a consequence, the plant's response to herbivore damage must integrate the responses to the herbivore-unspecific mechanical wounding and the herbivore-specific application of insect-derived chemical elicitors. Wound-induced resistance is to a large extent mediated by products of the octadecanoid pathway, which includes linolenic acid-derived compounds, such as 12-oxophytodienoic acid, jasmonic acid and methyl jasmonate (Creelman and Mullet 1997; Wasternack and Parthier 1997). However, at least two more signalling pathways, to ethylene and salicylic acid, are involved in the plant response to herbivores. Although it is becoming increasingly clear that single signal cascades, such as the oxylipins, can alone produce a bewildering array of potential secondary signal molecules with a diversity of functions (Creelman and Mullet 1997; Farmer et al. 1998; Wasternack and Parthier 1997), it has also become apparent that herbivore attack frequently involves the recruitment of several signalling cascades. The interaction between these different signalling pathways, widely referred to as 'signalling crosstalk', may explain the specificity of responses. Reymond and Farmer (1998) proposed a tuneable dial as a model for the regulation of defensive gene expression based on the crosstalk of the three signal pathways for jasmonic acid, salicylic acid and ethylene. How the responses are fine-tuned to optimize the defence against particular herbivore species or the attack by multiple species or guilds is the subject of a series of recent investigations (Bostock et al. 2001; Walling 2000; Thaler and Bostock

2004). Genoud and Metraux (1999) summarized examples of crosstalks between different signal pathways and modelled them as Boolean networks with logical linkages and circuits. The model complements earlier crosstalk models and makes concrete predictions regarding the outcome of the interactions between different signalling pathways. Currently such models are limited by our incomplete understanding of all the signalling cascades that are involved and sketchy knowledge about the biochemical consequences of the expression and interactions of these pathways. Also sketchy is the understanding of how signal crosstalk translates to ecological interactions among players of the second and the third trophic levels and how compromised plant defence responses translate into plant fitness and eventually influence the coevolutionary process between plants and insects. An understanding of the functional consequences of signal crosstalk and the resulting expression of the various plant defences requires a sophisticated understanding of the whole plant function and natural history of the involved multitrophic interaction networks in the plants' native habitats (Kessler et al. 2004; Steppuhn et al. 2004).

The wild tobacco plant *Nicotiana attenuata* (Torr.ex Watts) is a study system in which modern molecular and chemical-analytical tools are being applied in field and laboratory experiments to understand the complex plant–insect interactions. The system, propagated by Ian T. Baldwin and his co-workers at the Max Planck Institute for Chemical Ecology in Jena, Germany, is a prime example of the modern consolidation of different research domains. In the following paragraph I will give a brief introduction into the study system and highlight studies that illustrate the complexity of species interactions that result from herbivore-induced plant responses and the potential importance of inducible plant defences for structuring the plant's arthropod community.

THE WILD TOBACCO NICOTIANA ATTENUATA

The wild tobacco plant *N. attenuata* grows ephemerally in Great Basin desert habitats in the southwestern USA. It germinates from long-lived seed banks in response to chemical cues in wood smoke (Preston and Baldwin 1999). One such compound, the butenolide 3-methyl-2*H*-furo[2, 3-*c*]pyran-2-one, has recently been identified and found to promote seed germination in a number of plant species (Flematti et al. 2004). The 'fire-chasing behaviour' of *N. attenuata* forces the plant's arthropod herbivore community to re-establish with every new plant population. Inducible plant defences are thought to be an adaptation to such unpredictable herbivore pressure (Karban and Baldwin 1997). Wild tobacco increases its production of secondary metabolites (nicotine, phenolics, diterpeneglycosides, VOCs) and defensive proteins (trypsin proteinase inhibitors (TPI)) after attack by herbivores such as *Manduca* hornworms, *Tupiocoris notatus* bugs or *Epitrix hirtipennis* beetles (Kessler and Baldwin 2001; 2004), as well as in response to mechanical damage, or by elicitation with methyl jasmonate (Halitschke et al. 2000; Keinanen et al. 2001; Van Dam and Baldwin 2001).

Although the responses to these different elicitors frequently differ qualitatively and quantitatively, they diminish the plant's palatability to herbivores (direct defence) (Steppuhn et al. 2004; Van Dam et al. 2000; Zavala et al. 2004b) and/or increase its attractiveness to the natural enemies of the herbivores (indirect defence) (Kessler and Baldwin 2001; 2004). Anti-digestive proteins such as TPIs are known from several plant species to play a direct defensive role (Koiwa et al. 1997; Tamayo et al. 2000). Recent studies with natural mutants and antisense-transformed *N. attenuata* plants that are deficient in the induced production of TPIs, provide striking evidence for the defensive function of these anti-digestive enzymes. *Manduca sexta* caterpillars grow significantly faster and suffer from lower mortality rates on TPI-deficient plants than on plants with an intact TPI response or on plants that constitutively produced TPIs (Zavala et al. 2004b). The production of TPIs results in significant physiological fitness costs for the plant (Zavala et al. 2004a). In nature, the induction of plant defences in the absence of herbivores causes a significant reduction in lifetime seed production (Baldwin 1998). In contrast to the nitrogen-consuming production of defences such as nicotine or TPIs, the herbivore-induced production of VOCs is thought to be less costly (Halitschke et al. 2000). However, their indirect defensive effects may be not less important for the plant's fitness.

N. attenuata produces a series of VOCs, which derive from at least three different biochemical pathways (terpenoids, oxylipins, shikimates), in response to herbivore damage (Halitschke et al. 2000; Kessler and Baldwin 2001). Interestingly the four herbivore species (*M. sexta*, *M. quinquemaculata*, *T. notatus* and *E. hirtipennis*) that had been used in experiments elicited the emission of similar VOCs from *N. attenuata* plants (Kessler and Baldwin 2001). However, the quantities of the specific compounds, produced by the plant, differed significantly after the elicitation by different herbivore species. Some of the commonly emitted compounds have also been identified in the headspace of other plant species (Pare and Tumlinson 1998; Takabayashi and Dicke 1996; Turlings and Benrey 1998). Therefore it had been hypothesized that they may function as universal signs of herbivore damage and should, if singly emitted in the background of the plants' natural emissions, attract generalist predators in nature. The hypothesis proved right in that a generalist predator, the big-eyed bug *Geocoris pallens,* was attracted by the entire herbivore-induced VOCs bouquet as well as by single compounds (Kessler and Baldwin 2001; James 2005). In addition, adult *Manduca* moths used the same VOC signal to avoid already damaged plants for oviposition and thereby avoid increased predation pressure and reduced food quality as a result of induced direct defences. As a consequence, the multiplicative effect of the bottom-up and top-down components of herbivore-induced VOC emission was significant. It could reduce the numbers of *N. attenuata*'s most damaging herbivore, *M. quinquemaculata* by over 90% (Kessler and Baldwin 2001).

N. attenuata is attacked by many herbivore species from different feeding guilds in nature. However, these species may not always co-occur on the same plant due to plant-mediated effects. For example, the leaf-chewing larvae of the sympatric sibling species *M. sexta* and *M. quinquemaculata* tend not to co-occur with the sap-sucking mirid *T. notatus,* even when both species are found in adjoining host populations. Moreover, in plant populations with high numbers of plants infested by *T. notatus* the mortality of *Manduca* larvae and the seed-capsule production of

N. attenuata plants was higher than in plant populations without *T. notatus* (Kessler and Baldwin 2004). The apparent mechanism of this antagonistic relationship between two herbivore species and its fitness consequences for the plant reflects the complexity of plant–insect interactions and the size of the arena in which the interaction is played out (Figure 1).

That the two hornworm species and the mirid bugs seemed not to interact directly led us to hypothesize that plant-mediated effects caused the seemingly competitive interaction between the herbivores. Indeed, *M. sexta* and *M. quinquemaculata* hornworms grew much more slowly on plants that previously had been damaged by *T. notatus* than on undamaged plants. Interestingly, the metabolic responses to the damage by leaf-chewing hornworms and piercing-sucking mirids seemed very similar. The concentrations of a series of plant resistance-related secondary metabolites (phenolics and diterpene glycosides) and TPI were similarly increased in hornworm and mirid-damaged plants compared to undamaged plants. In confirmation with this result the *Manduca* larvae grew slower on plants that had been damaged by both conspecific caterpillars and mirids than on undamaged plants. Moreover, the emission of VOCs as well as the production of direct defensive compounds was increased after the damage by both herbivore species. Herbivore-induced VOCs in turn can function as indirect defences by attracting predators, such as *G. pallens,* to the damage site. As attack from both species elicits rather similar direct and indirect defensive plant responses, it was likely that the ecological context of these similar responses determines fitness consequences of the interaction for the *Manduca* hornworms and as a consequence for the plants (Kessler and Baldwin 2004).

One fitness benefit for the plant arises from the natural history of its interactions with herbivores. *Manduca* hornworms can consume three to five plants before they reach the pupal stage and therefore are considered the most damaging insect herbivores on *N. attenuata*. The hornworms usually depart before the plant is completely consumed, but the amount of leaf tissue lost to hornworm feeding is negatively correlated to the lifetime seed-capsule production of *N. attenuata*. Therefore, the plant's fitness costs from hornworm damage depend strongly on the developmental stage in which the hornworm leaves the plant or is removed by natural enemies such as parasitoids or predators. The growth-reducing effect of TPIs and secondary metabolites elicited by previous hornworm and mirid attack causes subsequently feeding hornworms to remain longer in the first two larval instars. As a consequence, the younger, more vulnerable hornworms are exposed longer to the dominating predator, the big-eyed bug *Geocoris pallens*, which is additionally attracted by the herbivore-induced VOCs (Kessler and Baldwin 2004). The direct effects of mirid-induced plant responses amplify the indirect defensive effects of predator attraction with negative fitness effects for the hornworms. Moreover, the predators prefer young hornworms over mirids as prey, which adds yet another factor contributing to the outcome of the interaction between the plant and its insect community.

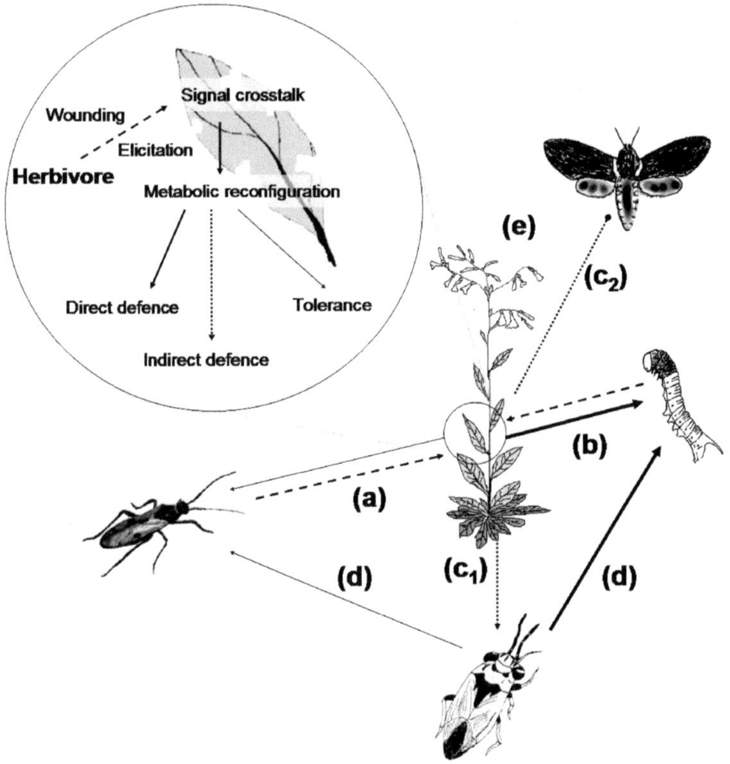

Figure 1. *The herbivorous mirid bug Tupiocoris notatus vaccinates the wild tobacco plant, Nicotiana attenuata against the more damaging tomato hornworm, Manduca quinquemaculata.* ***(a)*** *T. notatus damage (leaf-tissue wounding in combination with the application of salivary excretions) elicits a reconfiguration of the plant's secondary and primary metabolism.* ***(b)*** *The resulting mirid-induced production of toxic and anti-digestive plant compounds functions as* ***direct defence*** *and reduces the growth of the more damaging herbivore M. quinquemaculata, which therefore remains longer in the for predators vulnerable first two instars.* ***(c)*** *In addition the plant releases volatile organic compounds in response to mirid and hornworm damage, which attract the predatory bug Geocoris pallens to the plant (****indirect defence****) (c_1) and repel adult M. quinquemaculata moths from oviposition (c_2).* ***(d)*** *The predator G. pallens prefers young Manduca hornworms over Tupiocoris bugs as prey. The direct effects of mirid-induced plant responses amplify the indirect defensive effects of predator attraction with negative fitness effects for the hornworms.* ***(e)*** *Mirid-damaged plants, in contrast to hornworm-damaged plants, seem to compensate metabolically for the allocation of resources (****tolerance****) into defences and produce the same number of seeds as undamaged control plants. With the elicitation of induced direct and indirect responses and the neutral effect on plant fitness, T. notatus attack 'vaccinates' N. attenuata plants against the more severely damaging Manduca hornworms. Manduca damage also induces the production of toxic and anti-digestive plant compounds but results in a significant fitness loss for the plant. The effects of the herbivore-induced plant responses on Tupiocoris fitness remain unknown*

Interestingly, the reproductive consequences of hornworm and mirid attack are very different for the plant. While the plant metabolically responds very similarly to hornworm and mirid attack and gains resistance to hornworms, attack by mirids (in contrast to attack by hornworms) does not reduce the reproductive success of the plant, although the damage from these piercing-sucking insects can be substantial. Thus, mirid-damaged plants seem to compensate metabolically for the allocation of resources into defences. A differential display-reverse transcriptase PCR and subtractive library study of mirid-attacked *N. attenuata* plants (Voelckel and Baldwin 2003) revealed a series of mirid-specific transcriptional responses, which suggest that an adjustment of the primary metabolism is involved in the plant's ability to tolerate mirid attack. Particularly interesting is the mirid-induced increase in ribulose-1,5 bisphosphate carboxylase (RuBPCase) activase transcripts, which code for a stromal, regulatory protein that regulates the activity of the key enzyme in CO_2 assimilation, RuBPCase (Portis 1995). In addition, a cDNA microarray analysis that compared the transcription patterns induced by mirids and hornworms, respectively, identified that herbivore-specific changes occur largely in the primary metabolism and signalling cascades rather than secondary metabolism (Voelckel and Baldwin 2004). Experiments with *Datura wrightii* reported similar neutral effects of *T. notatus* attack on plant fitness and suggested that damage by *T. notatus* may reduce photosynthetic capacity less than equivalent damage by chewing insects does (Elle and Hare 2000; Hare and Elle 2002). With the elicitation of induced direct and indirect responses and the neutral effect on plant fitness, *T. notatus* attack literally 'vaccinates' *N. attenuata* plants against the more severely damaging *Manduca* hornworms. The neutral effects on plant fitness and herbivore-induced plant defences in the context of the particular life-history traits of the interacting species provide the mechanism for the plant vaccination phenomena (Figure 1). The study shows that a suite of rather similar responses to attack from different herbivores can result in dramatic differences in plant fitness and illustrates the importance of studying plant–insect interactions in the rough and tumble of the natural environment (Kessler and Baldwin 2004).

Herbivore-induced responses as the one described above depend to a great extent on a functioning oxylipin signalling in the plant. In *N. attenuata* the wounding of leaf tissue is recognized by an endogenous jasmonic-acid (JA) burst (Baldwin et al. 1997; Schittko et al. 2000) that results in the expression of a series of defence-related genes (Halitschke et al. 2001) and eventually in the increased production of defensive compounds such as nicotine and TPIs (Baldwin et al. 1997; Van Dam et al. 2000). However, the plant response to herbivory frequently differs from the response to mechanical damage of the leaf tissue. For example, the attack from *Manduca* caterpillars is recognized by the plant as evidenced by a JA burst that is far greater than that produced by mechanical wounding (Halitschke et al. 2001; Schittko et al. 2000). This JA burst is associated with the expression of wound-responsive and JA-independent genes, and the introduction of oral secretions from the feeding caterpillar account for the differences (Halitschke et al. 2003). Interestingly, the specific elicitation by caterpillar oral secretions accounts also for an ethylene burst (in addition to the JA burst) in response to herbivore damage, which attenuates the damage-induced accumulation of nicotine (Kahl et al. 2000). The ethylene burst antagonizes the wound-induced transcriptional increase in the nicotine biosynthetic genes NaPMT1 and NaPMT2 (Winz and Baldwin 2001).

Current experiments with genetically transformed *N. attenuata* plants in their native habitat emphasize the crucial role of oxylipin signalling for the plant's herbivore defence and the impact of induced plant defences on the arthropod community composition. Halitschke and co-workers (Halitschke and Baldwin 2003; Halitschke et al. 2004) generated transformed *N. attenuata* lines, which expressed *N. attenuata* lipoxygenase 3 (NaLOX3), hydroperoxide lyase (NaHPL) and allene oxide synthase (NaAOS) in an antisense orientation. All three enzymes are key regulators in two distinct oxylipin pathways and play a major role in the plant's wound recognition. In laboratory studies, plants deficient in the expression or recognition of octadecanoids, derived from LOX3, are unable to elicit defence compounds and are more susceptible to herbivore attack. The herbivore resistance can be restored by externally treating the LOX3-deficient plants with methyl jasmonate (the methyl ester of jasmonic acid) (Halitschke and Baldwin 2003). Interestingly, AOS-deficient *N. attenuata* plants partially reduced JA and defence-compound accumulation but this did not attenuate the resistance to herbivores, which was attributed to a leaky genotype and is currently under further investigation. HPL-deficient plants did not produce C_6-aldehydes and alcohols (green-leaf volatiles), which can function as defences (antimicrobial and as direct defences against some herbivores) or as wound signals to transmit information within (Sivasankar et al. 2000) and between plants (Arimura et al. 2000). However, HPL-deficient plants retained their resistance against hornworm damage despite their potential signal function. In fact hornworms in the laboratory consumed and grew more slowly on HPL-deficient plants than on wild-type control plants. The hornworms' growth rate could be restored to the levels of wild-type plants if GLVs were added to the HPL-deficient transformants, which suggests that GLVs stimulate feeding by *Manduca* hornworms (Halitschke et al. 2004).

The example of plant vaccination by mirid herbivores illustrates how important the ecological context is when interpreting the function of an interaction-mediating trait. Therefore we exposed the same transformants that so convincingly confirmed the crucial function of oxylipin signalling in the laboratory to the natural arthropod community in their native habitat in southwestern Utah (Kessler et al. 2004). First, the plants responded to standard bioassays with *Manduca* caterpillars in the field much as they did in the laboratory. The most pronounced effect was the loss of resistance in LOX3-deficient plants. The plants were more susceptible to *Manduca* damage in the standardized bioassay and received more damage from natural herbivory than wild-type control plants and the two other transformed plant lines. However, a more detailed analysis of the herbivore community that had established on the plants revealed that the herbivore-induced plant responses can alter the host spectrum of generalist herbivores. We found two new herbivore species on the LOX3-deficient plants that do not usually feed on *N. attenuata*: a leafhopper *Empoasca sp.* and the western cucumber beetle *Diabrotica undecimpunctata tenella* (Figure 2) In fact, most of the observed damage on LOX3-deficient

Figure 2. *New herbivores on lipoxygenase3 (LOX3)-deficient Nicotiana attenuata plants.* ***(a)*** *The leafhopper Empoasca sp. and* ***(b)*** *the leaf beetle Diabrotica undecimpunctata tenella do not usually feed on wild tobacco N. attenuata, but use plants, transformed with an antisense construct of N. attenuata LOX3 as new host plants in nature*

plants resulted from one of the new herbivores, Empoasca sp., which successfully reproduced on the new, undefended host plants. The results of this study demonstrated that the LOX3-mediated inducibility of plant responses is crucial for the oviposition decision and the opportunistic host selection behaviour of generalist herbivores such as *Empoasca* sp. and *D. undecimpunctata*. Host selection thus seems determined not only by the plant's constitutively expressed chemical phenotype and external mortality factors but also by the plant's ability to induce responses to herbivory (Kessler et al. 2004). As with the discovery of the plant vaccination effect of mirid damage, the study with plants that are not able to induce responses to herbivory emphasizes the role of induced plant defences in structuring arthropod communities. Moreover, the few selected examples from the *N. attenuata* system point to the value both of using genetically silenced plants and molecular tools in ecological research and of studying plant–insect interactions in the full complexity of the natural environment.

PLANT–INSECT INTERACTIONS AND GENOME PROJECTS – A CONCLUSION

Genomic and molecular technologies have expanded the types of questions that can be addressed in the research of plant–insect interactions and ecology as a whole. Modern genomic and molecular approaches provide ways to examine physiological mechanisms of biological interactions including elicitation of responses, signal perception and transduction by the plant at the cellular level, and the ecological function of traits, such as the fitness effects of plant defences on the whole-plant and community level. However, ecologists interested in using genomic tools are currently restricted to the limited number of model organisms that already have significant genomic resources available (Thomas and Klaper 2004). The development of genomic tools for research on ecological study systems with well characterized natural histories appears to be too time- and resource-consuming to be achieved in our rather ephemeral and resource-limited research environments. On the other hand ecologists have only begun to utilize the already available genetic and molecular model systems to answer ecological questions although the current success of this approach is promising. For example, the list of secondary metabolites isolated from the genetic model plant *A. thaliana* has grown more than five-fold in the last ten years and the biosynthetic pathways resulting in these compounds as well as their ecological function are revealed with breathtaking speed (D'Auria and Gershenzon 2005). The vast diversity of available *Arabidopsis* mutants and the applicability of the developed genetic tools for studies on related species have inspired a number of ecological studies (e.g., Clauss et al. 2002; Cipollini et al. 2003; Van Poecke and Dicke 2004; Cipollini et al. 2005) which provide a basic building block for future research in the study of species interactions.

The constantly growing number of plant species whose genome will be partially or fully sequenced will allow these tools also to be applied to ecological (supplementary to genetic) model systems in two ways: a) the increasing number of genetic model systems also increases the number of wild relatives to which the developed tools can readily be applied; b) the development of genetic tools for new systems will be faster and more cost-efficient. For example, comparative genomics provides a tool to utilize the increasing sequence information from model plant species to clone genes that mediate the plants' resistance to herbivores from less studied native species (Mueller et al. 2005). Comparative approaches have already been used in modern plant breeding to identify genes that are involved in plant development and resistance to abiotic and biotic stresses (King 2002; Shimamoto and Kyozuka 2002). In addition, manipulative techniques such as genetic transformation methods can help to reveal the function and ecological relevance of defensive traits in nature (Kessler et al. 2004; Steppuhn et al. 2004).

The recently launched Solanaceae Genome project, although focusing on the genome sequence of the domesticated tomato Lycopersicon esculentum (*Solanum lycopersicon*), promotes the parallel sequencing and comparative biology of a number of species in the Solanaceae family, including wild species (www.sgn.cornell.edu). That way it will supplement and extend the opportunities given by the classical genetic model plants and increase the number of potential systems to study multi-species interactions in nature. Utilizing the new genetic tools and information and apply them in native plant systems to answer ecological and evolutionary questions will be crucial to understand the mechanisms of species interactions. And, it is through this integrative approach that we will be enabled to reveal how cells, organisms and ecosystems function.

With the growing appreciation of the importance of species interactions in natural as well as in agricultural systems, the success of the genome projects will be increasingly measured by their contributions to integrative biological research fields. Therefore, the modern consolidation of the once-separated biological research domains becomes a research necessity as well as a logical consequence of these domains' conceptual interdependence. The *N. attenuata* example nicely illustrates the multiple spatial scales on which plant–insect interactions are played out. In addition it emphasizes both the value of using genetic and molecular tools in ecological research and, more importantly, the value of profound natural-history knowledge when studying multi-species interactions. *N. attenuata* is only one out of the estimated 230-422,000 flowering plant species interacting with only a few of the estimated 2 to 30 million insect species. In order to understand the patterns in community ecology and biodiversity we may not need to study all the possible interactions. But in order to apply our knowledge in agriculture and species conservation we will need at least a few well studied examples derived from a good number of different habitats. In short, and most importantly: we must never stop exploring in the old naturalist's way.

ACKNOWLEDGEMENTS

I thank Paul Feeny, Anurag A. Agrawal and Jennifer S. Thaler for helpful comments on an earlier draft and Ian T. Baldwin for supporting and supervising the highlighted studies on the wild tobacco *N. attenuata* and promoting the valuable discussion of the usefulness of transformed plants in ecological research.

REFERENCES

Adler, L.S., Karban, R. and Strauss, S.Y., 2001. Direct and indirect effects of alkaloids on plant fitness via herbivory and pollination. *Ecology,* 82 (7), 2032-2044.

Agrawal, A.A., 1998. Induced responses to herbivory and increased plant performance. *Science,* 279 (5354), 1201-1202.

Agrawal, A.A., 2000. Overcompensation of plants in response to herbivory and the by-product benefits of mutualism. *Trends in Plant Science,* 5 (7), 309-313.

Agrawal, A.A., Strauss, S.Y. and Stout, M.J., 1999. Costs of induced responses and tolerance to herbivory in male and female fitness components of wild radish. *Evolution,* 53 (4), 1093-1104.

Alborn, H.T., Turlings, T.C.J., Jones, T.H., et al. 1997. An elicitor of plant volatiles from beet armyworm oral secretion. *Science,* 276 (5314), 945-949.

Arimura, G.I., Ozawa, R., Shimoda, T., et al. 2000. Herbivory-induced volatiles elicit defence genes in lima bean leaves. *Nature,* 406 (6795), 512-515.

Baldwin, I.T., 1998. Jasmonate-induced responses are costly but benefit plants under attack in native populations. *Proceedings of the National Academy of Sciences of the United States of America,* 95 (14), 8113-8118.

Baldwin, I.T., 2001. An ecologically motivated analysis of plant-herbivore interactions in native tobacco. *Plant Physiology,* 127 (4), 1449-1458.

Baldwin, I.T., Zhang, Z.P., Diab, N., et al. 1997. Quantification, correlations and manipulations of wound-induced changes in jasmonic acid and nicotine in *Nicotiana sylvestris. Planta,* 201 (4), 397-404.

Barbosa, P., Gross, P. and Kemper, J., 1991. Influence of plant allelochemicals on the tobacco hornworm and its parasitoid *Cotesia congregata. Ecology,* 72 (5), 1567-1575.

Berenbaum, M.R., 2002. Postgenomic chemical ecology: from genetic code to ecological interactions. *Journal of Chemical Ecology,* 28 (5), 873-896.

Bostock, R.M., Karban, R., Thaler, J.S., et al. 2001. Signal interactions in induced resistance to pathogens and insect herbivores. *European Journal of Plant Pathology,* 107 (1), 103-111.

Campbell, B.C. and Duffey, S.S., 1981. Alleviation of alpha -tomatine-induced toxicity to the parasitoid, *Hyposoter exiguae*, by phytosterols in the diet of the host, *Heliothis zea. Journal of Chemical Ecology,* 7 (6), 927-946.

Cipollini, D., Mbagwu, J., Barto, K., et al. 2005. Expression of constitutive and inducible chemical defenses in native and invasive populations of *Alliaria petiolata. Journal of Chemical Ecology,* 31 (6), 1255-1267.

Cipollini, D.F., Busch, J.W., Stowe, K.A., et al. 2003. Genetic variation and relationships of constitutive and herbivore-induced glucosinolates, trypsin inhibitors, and herbivore resistance in *Brassica rapa. Journal of Chemical Ecology,* 29 (2), 285-302.

Clauss, M.J., Cobban, H. and Mitchell-Olds, T., 2002. Cross-species microsatellite markers for elucidating population genetic structure in *Arabidopsis* and *Arabis* (Brassicaeae). *Molecular Ecology,* 11 (3), 591-601.

Creelman, R.A. and Mullet, J.E., 1997. Biosynthesis and action of jasmonates in plants. *Annual Review of Plant Physiology and Plant Molecular Biology,* 48, 355-381.

D'Auria, J.C. and Gershenzon, J., 2005. The secondary metabolism of *Arabidopsis thaliana*: growing like a weed. *Current Opinion in Plant Biology,* 8 (3), 308-316.

De Moraes, C.M., Mescher, M.C. and Tumlinson, J.H., 2001. Caterpillar-induced nocturnal plant volatiles repel conspecific females. *Nature,* 410 (6828), 577-580.

Dicke, M. and Hilker, M., 2003. Induced plant defences: from molecular biology to evolutionary ecology. *Basic and Applied Ecology,* 4 (1), 3-14.

Dicke, M. and Van Loon, J.J.A., 2000. Multitrophic effects of herbivore-induced plant volatiles in an evolutionary context. *Entomologia Experimentalis et Applicata,* 97 (3), 237-249.

Doss, R.P., Oliver, J.E., Proebsting, W.M., et al. 2000. Bruchins: insect-derived plant regulators that stimulate neoplasm formation. *Proceedings of the National Academy of Sciences of the United States of America,* 97 (11), 6218-6223.

Duffey, S.S. and Stout, M.J., 1996. Antinutritive and toxic components of plant defense against insects. *Archives of Insect Biochemistry and Physiology,* 32 (1), 3-37.

Ehrlich, P.R. and Raven, P.H., 1964. Butterflies and plants: a study in coevolution. *Evolution,* 18 (4), 586-608.

Eisner, T. and Berenbaum, M., 2002. Chemical ecology: missed opportunities? *Science,* 295 (5562), 1973.

Elle, E. and Hare, J.D., 2000. No benefit of glandular trichome production in natural populations of *Datura wrightii*? *Oecologia,* 123 (1), 57-65.

Farmer, E.E., Weber, H. and Vollenweider, S., 1998. Fatty acid signaling in *Arabidopsis*. *Planta,* 206 (2), 167-174.

Felton, G.W. and Eichenseer, H., 1999. Herbivore saliva and its effects on plant defense against herbivores and pathogens. *In:* Agrawal, A.A., Tuzun, S. and Bent, E. eds. *Induced plant defenses against pathogens and herbivores: biochemistry, ecology, and agriculture*. APS Press, St. Paul, 19-36.

Fine, P.V.A., Mesones, I. and Coley, P.D., 2004. Herbivores promote habitat specialization by trees in Amazonian forests. *Science,* 305 (5684), 663-665.

Flematti, G.R., Ghisalberti, E.L., Dixon, K.W., et al. 2004. A compound from smoke that promotes seed germination. *Science,* 305 (5686), 977.

Friedman, M.H., 2001. Traditional engineering in the biological century: the biotraditional engineer. *Journal of Biomechanical Engineering,* 123 (6), 525-527.

Genoud, T. and Metraux, J.P., 1999. Crosstalk in plant cell signaling: structure and function of the genetic network. *Trends in Plant Science,* 4 (12), 503-507.

Greene, H.W., 2005. Organisms in nature as a central focus for biology. *Trends in Ecology and Evolution,* 20 (1), 23-27.

Halitschke, R. and Baldwin, I.T., 2003. Antisense LOX expression increases herbivore performance by decreasing defense responses and inhibiting growth-related transcriptional reorganization in *Nicotiana attenuata*. *Plant Journal,* 36 (6), 794-807.

Halitschke, R., Gase, K., Hui, D.Q., et al. 2003. Molecular interactions between the specialist herbivore *Manduca sexta* (Lepidoptera, Sphingidae) and its natural host *Nicotiana attenuata*. VI. Microarray analysis reveals that most herbivore-specific transcriptional changes are mediated by fatty acid-amino acid conjugates. *Plant Physiology,* 131 (4), 1894-1902.

Halitschke, R., Kessler, A., Kahl, J., et al. 2000. Ecophysiological comparison of direct and indirect defenses in *Nicotiana attenuata*. *Oecologia,* 124 (3), 408-417.

Halitschke, R., Schittko, U., Pohnert, G., et al. 2001. Molecular interactions between the specialist herbivore *Manduca sexta* (Lepidoptera, Sphingidae) and its natural host *Nicotiana attenuata*. III. Fatty acid-amino acid conjugates in herbivore oral secretions are necessary and sufficient for herbivore-specific plant responses. *Plant Physiology,* 125 (2), 711-717.

Halitschke, R., Ziegler, J., Keinanen, M., et al. 2004. Silencing of hydroperoxide lyase and allene oxide synthase reveals substrate and defense signaling crosstalk in *Nicotiana attenuata*. *Plant Journal,* 40 (1), 35-46.

Hare, J.D. and Elle, E., 2002. Variable impact of diverse insect herbivores on dimorphic *Datura wrightii*. *Ecology,* 83 (10), 2711-2720.

Hartmann, T., 2004. Plant-derived secondary metabolites as defensive chemicals in herbivorous insects: a case study in chemical ecology. *Planta,* 219 (1), 1-4.

Harvey, J.A., Van Dam, N.M. and Gols, R., 2003. Interactions over four trophic levels: foodplant quality affects development of a hyperparasitoid as mediated through a herbivore and its primary parasitoid. *Journal of Animal Ecology,* 72 (3), 520-531.

Heil, M. and Baldwin, I.T., 2002. Fitness costs of induced resistance: emerging experimental support for a slippery concept. *Trends in Plant Science,* 7 (2), 61-67.

James, D.G., 2005. Further field evaluation of synthetic herbivore-induced plant volatiles as attractants for beneficial insects. *Journal of Chemical Ecology,* 31 (3), 481-495.

Kafatos, F.C. and Eisner, T., 2004. Unification in the century of biology. *Science,* 303 (5662), 1257.

Kahl, J., Siemens, D.H., Aerts, R.J., et al. 2000. Herbivore-induced ethylene suppresses a direct defense but not a putative indirect defense against an adapted herbivore. *Planta,* 210 (2), 336-342.

Karban, R. and Agrawal, A.A., 2002. Herbivore offense. *Annual Review of Ecology and Systematics,* 33, 641-664.

Karban, R. and Baldwin, I.T., 1997. *Induced responses to herbivory*. University of Chicago Press, Chicago.

Karban, R., Baldwin, I.T., Baxter, K.J., et al. 2000. Communication between plants: induced resistance in wild tobacco plants following clipping of neighboring sagebrush. *Oecologia,* 125 (1), 66-71.

Keinanen, M., Oldham, N.J. and Baldwin, I.T., 2001. Rapid HPLC screening of jasmonate-induced increases in tobacco alkaloids, phenolics, and diterpene glycosides in *Nicotiana attenuata*. *Journal of Agricultural and Food Chemistry,* 49 (8), 3553-3558.

Kessler, A. and Baldwin, I.T., 2001. Defensive function of herbivore-induced plant volatile emissions in nature. *Science,* 291 (5511), 2141-2144.

Kessler, A. and Baldwin, I.T., 2002. Plant responses to insect herbivory: the emerging molecular analysis. *Annual Review of Plant Biology,* 53, 299-328.

Kessler, A. and Baldwin, I.T., 2004. Herbivore-induced plant vaccination. Part I. The orchestration of plant defenses in nature and their fitness consequences in the wild tobacco *Nicotiana attenuata*. *Plant Journal,* 38 (4), 639-649.

Kessler, A., Halitschke, R. and Baldwin, I.T., 2004. Silencing the jasmonate cascade: induced plant defenses and insect populations. *Science,* 305 (5684), 665-668.

Khan, Z.R., Pickett, J.A., Van den Berg, J., et al. 2000. Exploiting chemical ecology and species diversity: stem borer and striga control for maize and sorghum in Africa. *Pest Management Science,* 56 (11), 957-962.

King, G.J., 2002. Through a genome, darkly: comparative analysis of plant chromosomal DNA. *Plant Molecular Biology,* 48 (1/2), 5-20.

Koiwa, H., Bressan, R.A. and Hasegawa, P.M., 1997. Regulation of protease inhibitors and plant defense. *Trends in Plant Science,* 2 (10), 379-384.

Lam, E., Kato, N. and Lawton, M., 2001. Programmed cell death, mitochondria and the plant hypersensitive response. *Nature,* 411 (6839), 848-853.

Mattiacci, L., Dicke, M. and Posthumus, M.A., 1995. Beta-glucosidase: an elicitor of herbivore-induced plant odor that attracts host-searching parasitic wasps. *Proceedings of the National Academy of Sciences of the United States of America,* 92 (6), 2036-2040.

Mueller, L.A., Solow, T.H., Taylor, N., et al. 2005. The SOL Genomics Network: a comparative resource for Solanaceae biology and beyond. *Plant Physiology,* 138 (3), 1310-1317.

Orr, D.B. and Boethel, D.J., 1986. Influence of plant antibiosis through 4 trophic levels. *Oecologia,* 70 (2), 242-249.

Pare, P.W. and Tumlinson, J.H., 1998. Cotton volatiles synthesized and released distal to the site of insect damage. *Phytochemistry,* 47 (4), 521-526.

Pohnert, G., Jung, V., Haukioja, E., et al. 1999. New fatty acid amides from regurgitant of lepidopteran (Noctuidae, Geometridae) caterpillars. *Tetrahedron,* 55 (37), 11275-11280.

Portis, A.R., 1995. The regulation of rubisco by rubisco activase. *Journal of Experimental Botany,* 46, 285-1291.

Preston, C.A. and Baldwin, I.T., 1999. Positive and negative signals regulate germination in the post-fire annual, *Nicotiana attenuata*. *Ecology,* 80 (2), 481-494.

Reddy, G.V.P. and Guerrero, A., 2004. Interactions of insect pheromones and plant semiochemicals. *Trends in Plant Science,* 9 (5), 253-261.

Reymond, P. and Farmer, E.E., 1998. Jasmonate and salicylate as global signals for defense gene expression. *Current Opinion in Plant Biology,* 1 (5), 404-411.

Schittko, U., Preston, C.A. and Baldwin, I.T., 2000. Eating the evidence? *Manduca sexta* larvae can not disrupt specific jasmonate induction in *Nicotiana attenuata* by rapid consumption. *Planta,* 210 (2), 343-346.

Shimamoto, K. and Kyozuka, J., 2002. Rice as a model for comparative genomics of plants. *Annual Review of Plant Biology,* 53, 399-419.

Sivasankar, S., Sheldrick, B. and Rothstein, S.J., 2000. Expression of allene oxide synthase determines defense gene activation in tomato. *Plant Physiology,* 122 (4), 1335-1342.

Sopow, S.L., Shorthouse, J.D., Strong, W., et al. 2003. Evidence for long-distance, chemical gall induction by an insect. *Ecology Letters,* 6 (2), 102-105.

Steppuhn, A., Gase, K., Krock, B., et al. 2004. Nicotine's defensive function in nature. *PLoS Biology,* 2 (8), 1074-1080.

Tabashnik, B.E., Liu, Y.B., Malvar, T., et al. 1998. Insect resistance to *Bacillus thuringiensis*: uniform or diverse? *Philosophical Transactions of the Royal Society of London. Series B. Biological Sciences,* 353 (1376), 1751-1756.

Takabayashi, J. and Dicke, M., 1996. Plant-carnivore mutualism through herbivore-induced carnivore attractants. *Trends in Plant Science,* 1 (4), 109-113.

Tamayo, M.C., Rufat, M., Bravo, J.M., et al. 2000. Accumulation of a maize proteinase inhibitor in response to wounding and insect feeding, and characterization of its activity toward digestive proteinases of *Spodoptera littoralis* larvae. *Planta,* 211 (1), 62-71.

Thaler, J.S. and Bostock, R.M., 2004. Interactions between abscisic-acid-mediated responses and plant resistance to pathogens and insects. *Ecology,* 85 (1), 48-58.

Thomas, M.A. and Klaper, R., 2004. Genomics for the ecological toolbox. *Trends in Ecology and Evolution,* 19 (8), 439-445.

Turlings, T.C.J. and Benrey, B., 1998. Effects of plant metabolites on the behavior and development of parasitic wasps. *Ecoscience,* 5 (3), 321-333.

Van Breusegem, F., Vranova, E., Dat, J.F., et al. 2001. The role of active oxygen species in plant signal transduction. *Plant Science,* 161 (3), 405-414.

Van Dam, N.M. and Baldwin, I.T., 2001. Competition mediates costs of jasmonate-induced defences, nitrogen acquisition and transgenerational plasticity in *Nicotiana attenuata. Functional Ecology,* 15 (3), 406-415.

Van Dam, N.M., Hadwich, K. and Baldwin, I.T., 2000. Induced responses in *Nicotiana attenuata* affect behaviour and growth of the specialist herbivore *Manduca sexta. Oecologia,* 122 (3), 371-379.

Van Poecke, R.M.P. and Dicke, M., 2004. Indirect defence of plants against herbivores: using *Arabidopsis thaliana* as a model plant. *Plant Biology,* 6 (4), 387-401.

Voelckel, C. and Baldwin, I.T., 2003. Detecting herbivore-specific transcriptional responses in plants with multiple DDRT-PCR and subtractive library procedures. *Physiologia Plantarum,* 118 (2), 240-252.

Voelckel, C. and Baldwin, I.T., 2004. Herbivore-induced plant vaccination. Part II. Array-studies reveal the transience of herbivore-specific transcriptional imprints and a distinct imprint from stress combinations. *Plant Journal,* 38 (4), 650-663.

Walling, L.L., 2000. The myriad plant responses to herbivores. *Journal of Plant Growth Regulation,* 19 (2), 195-216.

Wasternack, C. and Parthier, B., 1997. Jasmonate-signalled plant gene expression. *Trends in Plant Science,* 2 (8), 302-307.

Winz, R.A. and Baldwin, I.T., 2001. Molecular interactions between the specialist herbivore *Manduca sexta* (Lepidoptera, Sphingidae) and its natural host *Nicotiana attenuata.* IV. Insect-induced ethylene reduces jasmonate-induced nicotine accumulation by regulating putrescine N-methyltransferase transcripts. *Plant Physiology,* 125 (4), 2189-2202.

Zavala, J.A., Patankar, A.G., Gase, K., et al. 2004a. Constitutive and inducible trypsin proteinase inhibitor production incurs large fitness costs in *Nicotiana attenuata. Proceedings of the National Academy of Sciences of the United States of America,* 101 (6), 1607-1612.

Zavala, J.A., Patankar, A.G., Gase, K., et al. 2004b. Manipulation of endogenous trypsin proteinase inhibitor production in *Nicotiana attenuata* demonstrates their function as antiherbivore defenses. *Plant Physiology,* 134 (3), 1181-1190.

CHAPTER 4

THE EFFECT OF HOST-ROOT-DERIVED CHEMICAL SIGNALS ON THE GERMINATION OF PARASITIC PLANTS

RADOSLAVA MATÚŠOVÁ AND HARRO J. BOUWMEESTER[#]

[#] *Corresponding author, Plant Research International, P.O. Box 16, 6700 AA Wageningen, The Netherlands. E-mail: harro.bouwmeester@wur.nl*

Abstract. The parasitic plants *Orobanche* and *Striga* spp. are holo- and hemiparasites, which largely depend on a host plant to obtain their nutrients and water. The seeds of these parasites can only germinate in the presence of a chemical compound that is exuded from the roots of their host. These compounds are called germination stimulants and so far several of these compounds have been identified in the exudates of hosts (and false hosts) of several *Orobanche* and *Striga* species. The germination stimulants play an important role in fine-tuning of the lifecycle of the parasites to that of their hosts. In this chapter we describe the processes that play a role in this interaction, for example how the germination stimulants are produced by the host and how they are perceived by the parasite. Also we discuss the possible importance of the germination stimulants in determining host specificity.
Keywords: *Orobanche*; *Striga*; carotenoids; dormancy; host specificity; sensitivity

INTRODUCTION

Underground chemical signalling and parasitic plants

Chemical signalling between individuals of one species but also between individuals of different species plays an essential role in biology. Although plants cannot talk, listen or see, they communicate extensively, using secondary metabolites to convey messages (see Chapters 2 and 6; (Degenhardt et al. 2003; Dicke and Hilker 2003). Although the concept of communication of plants is perhaps less easy to imagine underground, underground signalling too is of great importance for plants (Bais et al. 2004). Examples are the colonization by nitrogen-fixing bacteria (rhizobia) and the attraction of insect-parasitic nematodes by insect-attacked roots (Limpens and Bisseling 2003; Rasmann et al. 2005). In all these signalling processes, the specificity of the interaction is very important and delicately regulated. Predators are attracted to plants attacked by their prey and rhizobia respond to the roots of legumes. In non-beneficial underground interactions chemical cues produced by the

M. Dicke and W. Takken (eds.), Chemical Ecology: From Gene to Ecosystem, 39-54.

host plant are also of great importance and also here the specificity is often amazing (Hirsch et al. 2003). An exciting example of plant–plant underground communication is the recognition by the parasitic plants *Orobanche* and *Striga* spp of chemical signals exuded by the roots of suitable host plants. The parasitic broomrapes and witchweeds can only survive on the roots of a host and must obtain most of their resources from them. The seeds of the parasitic plants are tiny, and after germination they must attach themselves to a host root within days or otherwise they will die (Butler 1995). Parasitic plants have evolved a graceful strategy to deal with this requirement: their germination depends unconditionally on compounds that are produced by the roots of their hosts in extremely low concentrations. These stimulants are collectively called strigolactones. The strigolactones belong to the chemical class of the isoprenoids, to which many of the known biologically active plant communication signals belong. Much is known about the biosynthesis of isoprenoids in above-ground plant organs; by contrast we know surprisingly little of this process in the root system. Until recently, the significance of the strigolactones for the plant itself has remained elusive (why do plants produce these compounds when they are obviously disadvantageous, since they cause parasitism?). The fact that they have persisted despite the supposedly strong counter-selection suggests that they are essential for the plant. Indeed, an intriguing recent study has shown that the strigolactones are used by arbuscular mycorrhizal fungi for their colonization process (the strigolactones are the branching factor that is required for mycorrhizal mycelia to become infective), and this most likely answers the question why plants still produce strigolactones (Akiyama et al. 2005; Matúšová et al. 2005).

Broomrapes (*Orobanche* spp.) and witchweeds (*Striga* spp.) (both *Scrophulariaceae*) can heavily infest crops with a large negative impact on agriculture in many countries. Orobanche spp. are holoparasites that are completely lacking chlorophyll and for their growth and development are completely dependent on their host for the supply of water and nutrients. *O. cumana* Wallr. parasitizes sunflower in eastern Europe around the Black Sea, in Spain (Akhtouch et al. 2002), and recently the pest was reported to spread widely in Israel (Aly et al. 2001). *O. ramosa* and *O. aegyptiaca* parasitize a wide range of hosts, such as tomato, potato, eggplant, tobacco, carrot, lettuce and many others (Press et al. 2001). *O. crenata* is a widespread parasite of legumes all around the Mediterranean (Press et al. 2001). *Striga* spp. belong to the hemiparasites with lower photosynthetic activity and basically behave as holoparasites (Parker and Riches 1993). They are serious pests in the African continent. Hosts of *S. hermonthica*, *S. asiatica*, *S. aspera* and *S. forbesii* include grain cereals such as maize, sorghum, millet and upland rice (Press et al. 2001). *S. gesnerioides* is a parasite of cowpea, and causes extensive damage in sub-Saharan dry areas, particularly West-Africa (Press et al. 2001).

Life cycle of Striga spp. and Orobanche spp.

The life cycles of *Striga* and *Orobanche* spp. are essentially similar; both start with the germination of the seed that is induced by compounds exuded by the roots of their hosts (Figure 1). After germination, the radicle grows towards the host root and forms a haustorium. The haustorium is formed by the swelling of the radicle tip with a hairy structure with which the parasite attaches itself to the host root (Hood et al. 1998). The establishment of a xylem connection, tubercule formation, shooting and seed production are the next steps in the life cycle (Figure 1). In many of these steps chemical communication occurs between the host plant and the parasite. This starts with the secretion of secondary metabolites from the roots of the host (and some non- or false hosts) that induce the germination of the seeds of the parasite. After germination, additional host-derived secondary metabolites play a role in the plant–parasite interaction. The orientation of the parasite's radicle growth towards the host root has been postulated to be directed by the concentration gradient of the germination stimulant (Dube and Olivier 2001) or by other host-root-derived compounds. Host-produced allelochemicals may interfere with the interaction between host and parasite. In sunflower, for example, coumarins were shown to be responsible for the inhibition of germination and necrosis of *O. cernua* after germination (Serghini et al. 2001). Attachment to the root of the host plant and the host–parasite xylem connection is mediated by a haustorium, of which the formation

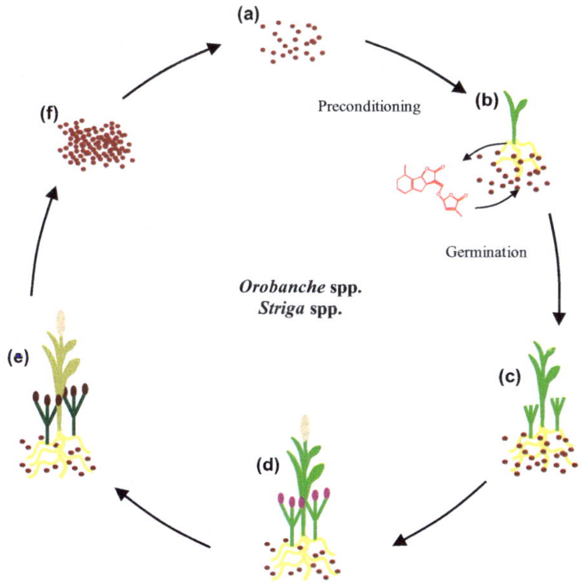

Figure 1. *Life cycle of parasitic plants Orobanche spp. and Striga spp. (a) the seeds are buried in the soil; (b) they become sensitive to the germination stimulants exuded by the roots of the host plant and may germinate; (c) the germinated seeds form a haustorium by which they attach themselves to the host root, establish a xylem connection and emerge; (d) parasitic plants flower; (e) they produce mature seeds and end up in a new generation of seeds in soil; (f) in the next season the cycle starts again (a)*

is initiated by host-derived secondary metabolites, notably phenolics (Keyes et al. 2001; Yoder 2001; Hirsch et al. 2003). Finally, after haustorium formation the penetration of intrusive cells into the host root xylem is realized, probably with the involvement of hydrolytic enzymes produced by the parasite (Losner-Goshen et al. 1998). Successful establishment of a xylem connection is also dependent on the host and can be terminated by host-produced toxins (Goldwasser et al. 1999; Labrousse et al. 2001; Serghini et al. 2001). Indeed, the resistance of some sorghum cultivars is based on the induction of necrosis at the attachment site on the root (Mohamed et al. 2003).

Germination stimulants

As described above the first involvement of chemical signalling in the life cycle of the parasitic plant is the induction of germination by germination stimulants. For *Striga* spp. several germination stimulants were identified from host and non-host plants. Most of them are known as strigolactones (Figure 2). The first identified germination stimulant was strigol; it was isolated from the non-host plant cotton (Cook et al. 1972). Recently, Yoneyama and co-workers isolated and characterized from cotton root exudates also strigyl acetate, which induces germination of *O. minor* (Sato et al. 2005). Germination stimulants in maize and sorghum were identified as strigol (Siame et al. 1993) and sorgolactone (Hauck et al. 1992). Alectrol was identified in the root exudate of cowpea (Muller et al. 1992). Alectrol and orobanchol were isolated and identified from the root exudate of red clover (Yokota et al. 1998) (Figure 2). The same group reported on the isolation of four novel strigolactones from the root exudate of tomato, and the presence of a novel strigol isomer in the root exudate of sorghum (Yoneyama et al. 2004). There are also several synthetic compounds inducing germination of parasitic plants (Reizelman and Zwanenburg 2002). Among them is the strigol analogue GR24, a very potent synthetic stimulant, which induces germination of many *Orobanche* and *Striga* spp. and is widely used as a positive control in most laboratory experiments (Figure 2).

It is obvious that the germination stimulants play a crucial role in the life cycle of parasitic plants and could also be an important target for the design of new control strategies for agriculturally important parasitic plants. Nevertheless, little is known about how these compounds are produced by the host, how they are perceived by the parasite and how selective this process of host recognition is. Here we will review our own work and that of others pertaining to these three subjects.

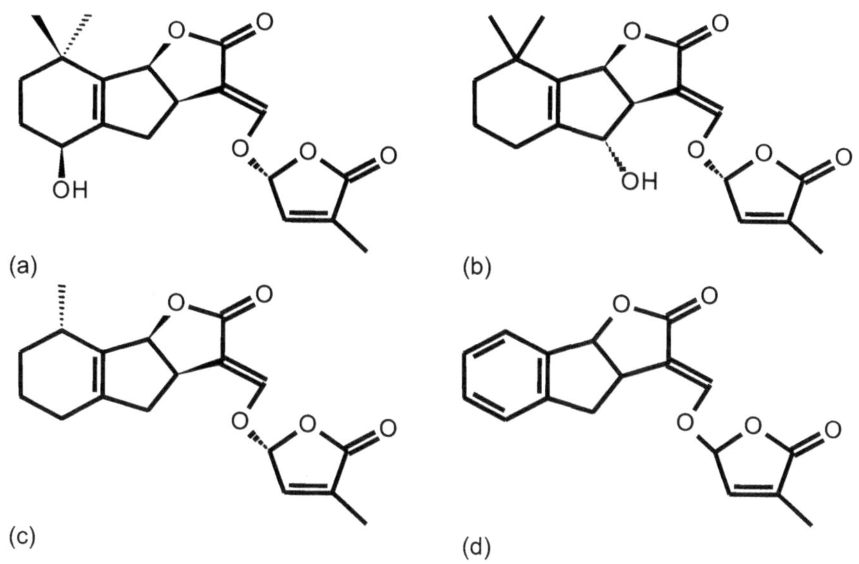

Figure 2. *Structure of strigolactone germination stimulants. (a) (+)-strigol; (b) orobanchol; (c) sorgolactone;(d) synthetic germination stimulant GR24*

PERCEPTION OF GERMINATION STIMULANTS

Germination stimulants are exuded from the roots of host plants in very low quantities. For example, seedlings of cotton produce about 14 pg of strigol per plant per day (Sato et al. 2005). Considering these extremely low amounts it is important that we are aware that the seeds used in studies on natural germination stimulants are sensitive to the stimulants and that this sensitivity is not a static parameter. Indeed, the availability of the synthetic germination stimulant GR24, of which in principle relatively large concentrations can be used (compared with the predicted concentrations of naturally occurring germination stimulants), has more or less obscured the importance (and variability) of sensitivity in a number of studies. To become responsive to the germination stimulants the seeds of *Orobanche* and *Striga* spp. require a moist environment for a certain period of time at a suitable temperature. This treatment is described as preconditioning or conditioning and is comparable to what is called (warm) stratification in seeds of non-parasitic plants or release of dormancy (Matúšová et al. 2004). During preconditioning, seeds become metabolically active (Mayer and Bar Nun 1997). The temperature used during preconditioning strongly affects the responsiveness to chemical stimulants. Seeds of *O. crenata* are able to germinate after preconditioning from 5°C to 30°C (Van Hezewijk et al. 1993). However, preconditioning at sub-optimal temperatures results in a lower sensitivity to the germination stimulant, which does not increase even after prolonged preconditioning (Van Hezewijk et al. 1993; Matúšová et al. 2004).

Preconditioning at an optimal temperature (e.g., about 20°C for *O. cumana* and 30°C for *S. hermonthica*) releases dormancy within 2-3 weeks and increases the sensitivity to GR24 by several orders of magnitude (Figure 3). After reaching maximum sensitivity, prolonged preconditioning induces secondary dormancy, i.e., decreased sensitivity of *O. cumana* and *S. hermonthica* to GR24 (Figure 3) (Matúšová et al. 2004). A similar trend was observed for *O. ramosa* (Gibot-Leclerc et al. 2004). It is important to note that the rapid changes in sensitivity during prolonged preconditioning are particularly visible at low concentrations of GR24. At higher concentrations, GR24 usually induces high germination, regardless of the preconditioning period. Parasitic plant seeds are highly sensitive to the germination stimulant for a short period of time only, and then enter into secondary dormancy relatively quickly. These large changes in sensitivity to germination stimulants are suggestive of a safety mechanism that ensures that seeds can respond to the germination stimulants produced by their host only during a restricted period of the year (assuming – and this is quite likely – that the hosts continue to produce strigolactones throughout further development). This is of great ecological importance as the parasitic plants require a long enough period of time to reproduce, and germination during the later stages of host development would not allow this. The similar pattern of increasing and decreasing sensitivity to GR24 that we observed with *S. hermonthica* seeds preconditioned for a prolonged period of time under field conditions suggests that the mechanism observed is indeed not just a laboratory phenomenon but is of ecological significance (Matúšová et al. 2004).

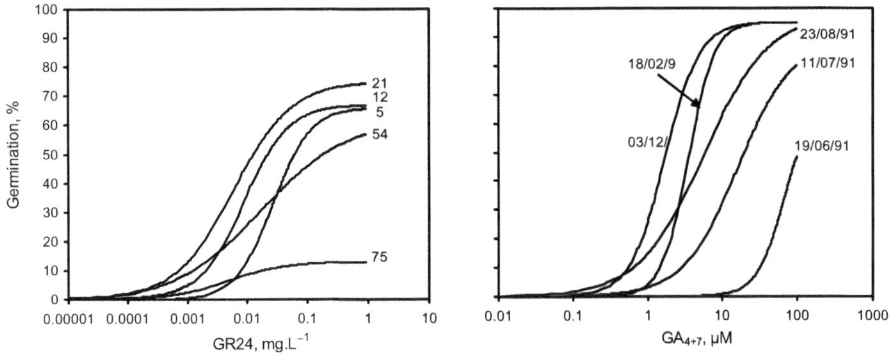

Figure 3. *A. Dose–response curves showing the effect of the preconditioning period on the sensitivity of Striga hermonthica to the germination stimulant GR24. Numbers indicate days of preconditioning at 30°C. B. Changes in gibberellin* GA_{4+7} *dose–response curves of Arabidopsis thaliana as a consequence of burial in the field. Dates indicate the date that seeds were exhumed and their germination tested in a range of gibberellin concentrations. Burial date: 19 June 1991 (Derkx and Karssen 1994)*

Interestingly, these changes in the sensitivity of the parasitic plant seeds to germination stimulants display a similarity to the dormancy (sensitivity) changes of seeds of non-parasitic wild plants to, for example, light (position), nitrate (growth conditions), and gibberellin (Derkx and Karssen 1993a; 1993b; 1994; Matúšová et al. 2004). For non-parasitic plants, this mechanism ensures that seeds will germinate and grow under favourable conditions only. Apparently, parasitic plants have adapted this mechanism to recognize suitable growing conditions also, i.e., the presence of a suitable host, by responding to a typical host/plant-produced metabolite. Indeed, the shift in the GR24-response curves and the shift in the gibberellin-response curves during dormancy relief in *Arabidopsis*, as reported by Derkx and Karssen (1994), are quite similar (Figure 3). Gibberellins and a putative gibberellin receptor play a crucial role in the germination of non-parasitic wild-plant seeds, even though changes in the sensitivity to gibberellins was hypothesized not to be the mechanism responsible for the changes in dormancy in the seeds of *Arabidopsis and Sisymbriun officinale* (Derkx and Karssen 1993a; 1994). According to a model proposed by Hilhorst and Karssen (1988), gibberellin biosynthesis and sensitivity to gibberellin in these seeds are controlled by a receptor that is activated by nitrate and red light (Hilhorst 1993; Hilhorst and Karssen 1988; Vleeshouwers et al. 1995). The structure of the strigolactone parasitic-plant germination stimulants and the gibberellins is fairly similar, and it is not unlikely that their respective receptors have a common origin (Matúšová et al. 2004). A gibberellin receptor in non-parasitic plant seeds was postulated by (Hilhorst et al. 1996; 1986). The involvement of a receptor in germination-stimulant recognition has been postulated (Wigchert and Zwanenburg 1999) and is supported by the dose–response curves (Figure 3) (Matúšová et al. 2004).

BIOSYNTHETIC ORIGIN OF GERMINATION STIMULANTS

Germination stimulants are exuded from the roots of host plants in very low concentrations, which makes the isolation and characterization of these compounds quite difficult. Moreover, the big losses during the isolation process and instability of these compounds (Sato et al. 2005) are reasons why large volumes of root exudates are still needed for their characterization. For the same reasons, also the study of the biosynthesis of these compounds is difficult. The strigolactone germination stimulants were isolated from a wide variety of plant species and induce germination of a range of parasitic plant species. Nevertheless, they are strikingly similar and are obviously derived from the same biosynthetic pathway. The strigolactones are usually defined to be sesquiterpene lactones (Butler 1995; Yokota et al. 1998), but there is also some structural similarity to higher-order terpenoids/isoprenoids such as abscisic acid and other compounds, which are derived from the carotenoid pathway (Parry and Horgan 1992; Tan et al. 1997; Boumeester et al. 2003).

Isoprenoids are biosynthesized from isopentenyl diphosphate (IPP) and the isomeric dimethylallyl diphosphate (DMAPP) via two independent pathways: the cytosolic mevalonic-acid (MVA) pathway and the plastidic, non-mevalonate, methylerythritol-phosphate (MEP) pathway. The plastidic MEP pathway produces IPP and DMAPP for the biosynthesis of monoterpenes, diterpenes, carotenoids, the plant hormones gibberellins and abscisic acid and the side chains of chlorophylls, plastoquinones and phylloquinones. Sesquiterpenes, sterols and triterpenes are produced from the cytosolic MVA pathway.

Elucidation of germination-stimulant biosynthetic pathway

To determine the biosynthetic origin of the germination stimulants produced by plants we used two approaches: (1) the use of specific inhibitors of isoprenoid pathways and (2) the use of defined mutants in predicted biosynthetic pathways. Inhibitors were applied to seedlings only during a number of days to ensure normal plant development. Because of the very low concentrations at which the germination stimulants are active an analytical method to study the consequences of our treatments on germination-stimulant formation could not be used. Instead we used a germination bioassay as a very sensitive and useful detection method to analyse production of germination stimulants even in single seedlings.

The isoprenoid-pathway inhibitors mevastatin (inhibitor of the cytosolic MVA pathway) and fosmidomycin (inhibitor of the plastidic MEP pathway) only had a minor effect on germination-stimulant formation, possibly because of the exchange of IPP that has been shown to occur between the two pathways, particularly upon the use of these inhibitors (Hemmerlin et al. 2003). However, the carotenoid-pathway inhibitor fluridone reduced root-exudate-induced germination by about 80% compared with control maize seedlings, suggesting that the germination stimulants produced by maize are derived from the carotenoid pathway (Figure 4) (Matúšová et al. in press). Therefore, we decided to analyse the induction of germination by a series of carotenoid mutants from the Maize Genetics COOP Stock Center, Urbana, Illinois. The root exudates of several maize carotenoid mutants *lw1, y10*, *al1*, *al1y3, vp5* and *y9* (Figure 4) were tested for induction of *S. hermonthica* seed germination. The seedlings of all mutants induced lower germination of *S. hermonthica* seeds in comparison to their corresponding wild-type phenotype siblings (Matúšová et al. in press). The carotenoid biosynthesis inhibitor fluridone blocks the activity of phytoene desaturase, which corresponds to the maize *vp5* locus (Li et al. 1996; Hable et al. 1998). Both fluridone-treated maize and the *vp5* mutant root exudates induced significantly lower germination of *S. hermonthica.* Also treatment with the herbicide amitrole that blocks lycopene cyclase in maize seedlings (Dalla Vecchia et al. 2001) resulted in lower germination of *S. hermonthica* seeds than induced by control seedlings. The results in germination bioassays with root exudates of amitrole-treated plants suggest that the germination stimulants are derived from the carotenoid pathway below lycopene (Figure 4) (Matúšová et al. in press).

Below this point in the carotenoid pathway there are unfortunately only few well-characterized mutants available and one putative inhibitor of the enzyme *9-cis*-epoxycarotenoid dioxygenase (NCED), naproxen (Lee and Milborrow 1997; Schwartz et al. 1997) (Figure 4). The formation of the germination stimulant of maize was reduced by the use of naproxen. Bioassays with maize *vp14*, a mutant of NCED, confirmed the result obtained with naproxen. Also *vp14* induced lower

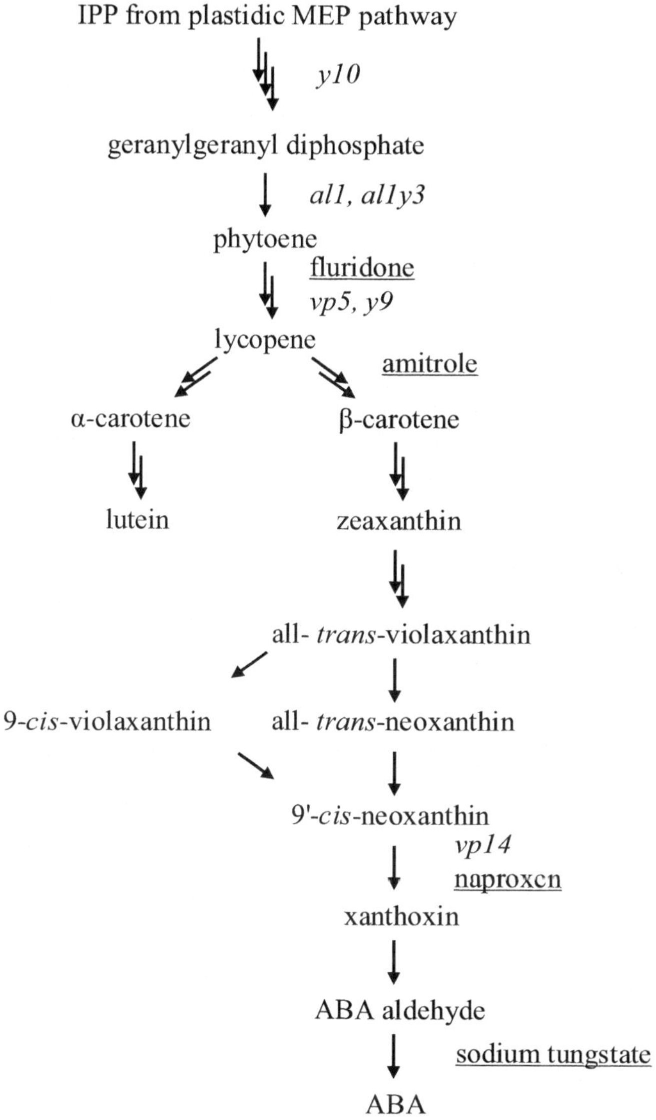

Figure 4. *Schematic representation of the carotenoid and abscisic-acid biosynthetic pathway. Carotenoid maize mutants (italics) and inhibitors (underlined) at different steps in the pathway are indicated*

germination. This suggests that carotenoid cleavage is involved in germination-stimulant biosynthesis, which is to be expected as the C40 carotenoids need to be cleaved in order to lead to the C14 (excluding the D-ring) strigolactones. The action of NCED leads to the formation of abscisic acid (ABA), and we tested whether ABA is a precursor of the germination stimulants. However, plants supplied with low concentrations of ABA induced much lower *S. hermonthica* germination, whereas treatment with the inhibitor of ABA-aldehyde oxidation, sodium tungstate, did not have any effect on *S. hermonthica* germination (Figure 4) (Matúšová et al. 2005). This shows that the germination stimulants are neither derived from intermediates below ABA aldehyde nor from ABA itself. The reduction of root-exudate-induced germination by ABA is most probably due to feedback inhibition by the exogenously applied ABA on the carotenoid pathway (Matúšová et al. 2005). In conclusion, the germination stimulants are derived from the carotenoid pathway through the action of a carotenoid-cleavage enzyme, possibly NCED. The cleavage may occur in several steps of the pathway and is expected to lead to the production of a C15 aldehyde, which we have postulated can be converted to the strigolactones in a number of enzymatic steps (Matúšová et al. 2005).

Germination of *S. hermonthica* is also induced by cowpea and sorghum root exudates (Gurney et al. 2002; Rugutt and Berner 1998). In cowpea root exudate the strigolactone alectrol has been identified (Muller et al. 1992), in sorghum exudates sorgolactone (Hauck et al. 1992) and hydroquinone (Chang et al. 1986). The root exudates of fluridone-treated cowpea induced about 80% less germination of *S. hermonthica* than those of non-treated cowpea. Interestingly, also the germination of *O. crenata* seeds with fluridone-treated-cowpea root exudate was less than that induced by the control. Fluridone treatment of sorghum seedlings almost completely blocked subsequent exudate-induced germination of *S. hermonthica* seeds (Matúšová et al. in press). These results show that the germination stimulant(s) of *S. hermonthica* exuded from the roots of cowpea and sorghum is (are) also derived from the carotenoid pathway. Also the cowpea-produced germination stimulant of *O. crenata* is derived from the carotenoid pathway. The germination stimulant(s) of *O. crenata* produced by its legume host(s) have not been identified yet, but our results suggest that this species also responds to a strigolactone germination stimulant. With regard to sorghum, Keyes and co-workers have claimed that the phenolic sorgoleone is the sorghum germination stimulant of *Striga* spp. (Keyes et al. 2001), but our results suggest that the natural sorghum germination stimulant is a strigolactone, such as sorgolactone. We have proven the carotenoid origin of germination stimulants for two parasitic plant species in three mono- and dicotyledonous hosts. At the same time, Yoneyama and co-workers have demonstrated strigolactones – known ones as well as new (tentatively) identified ones – in the root exudates of other plant species such as red clover and tomato (Yoneyama et al. 2004; Yokota et al. 1998), suggesting that carotenoid-derived germination-stimulant formation occurs in a variety of plant species.

ROLE OF GERMINATION STIMULANTS IN HOST SPECIFICITY

From the work by Yoneyama and coworkers (2004) on the identification of strigolactone germination stimulants it has become clear that there is a large structural diversity in the strigolactones (Yoneyama et al. 2004). Although the biological activity of the strigolactones resides mainly in the D ring (Mangnus and Zwanenburg 1992), an interesting question is whether the small changes in the remainder of the molecule have an effect on receptor binding in the parasitic plant seeds, and hence on host–parasite specificity. Of course host–parasite recognition/selectivity occurs at different stages of the life cycle also after germination (also see above). For example, the haustorial initiation and development up to attachment are very similar for host and non-host plants, but development following attachment differs for host (successful) and non-host (not successful) species (Hood et al. 1998). Nevertheless, the recognition of the germination stimulant is a crucial moment in the life cycle of the parasitic plants. Here, a strong selection pressure is present that should ensure that the seeds of the parasites only germinate in the presence of a true host and thus may complete their life cycle. Nevertheless, a number of examples suggest that the specificity may not be very high. Alectrol, for example, is inducing germination of *S. gesnerioides* (Muller et al. 1992), but it was also identified in red clover as a germination stimulant for *O.minor* (Yokota et al. 1998). Wigchert and Zwanenburg (1999) induced germination of the seeds of *O. crenata* – which normally parasitizes legumes – with sorgolactone, one of the germination stimulants identified in sorghum, and the root exudate of cowpea induces germination of *S. hermonthica*, which is known to parasitize monocotyledons. Finally, the synthetic strigolactone analogue GR24 (Figure 2) induces germination of many parasitic plant seeds regardless of parasite or host plant species.

On the other hand, there are examples of a certain degree of host specificity. Not all host plant species induce germination of all parasitic plant seeds. Also, not all synthetic germination stimulants induce germination of all parasites to the same extent (Mwakaboko 2003). We have compared the induction of germination of *S. hermonthica* batches collected from maize and sorghum by the exudates of maize (host), cowpea (non-host) and the synthetic germination stimulant GR24 (Table 1). Maize root exudates induced 36% germination of *S. hermonthica* seeds collected from maize. Cowpea root exudates induced 51% germination, and 0.001 $mg.l^{-1}$ of GR24 induced 44% germination of the same *S. hermonthica* seeds. The highest germination (62%) was induced by 0.1 $mg.l^{-1}$ GR24 (Table 1). In contrast, *S. hermonthica* seeds collected from sorghum germinated to 37% in response to the maize root exudate, to 22% in response to the cowpea exudate and to 49% in response to 0.001 $mg.l^{-1}$ of GR24, whereas the maximum germination in response to 0.1 $m.l^{-1}$ GR24 was 96%. *S. hermonthica* collected from another sorghum field responded to maize and cowpea root exudates by very low germination (15 and 14%, respectively), even though germination in 0.001 and 0.1 $mg.l^{-1}$ GR24 was as high as 28% and 89%, respectively (Table 1). The slightly different response of the two sorghum-collected *S. hermonthica* batches may be due to the fact that different sorghum varieties may have different root exudate compositions.

Table 1. *Germination of Striga hermonthica induced by root exudates of maize, cowpea and the synthetic germination stimulant GR24. Numbers are averages of 6 individual replicates ± SE).*

Origin of *Striga hermonthica* seeds	Germination (%) induced by				
			GR24, $mg.l^{-1}$		
	maize	cowpea	0.001	0.01	0.1
Maize (Kenya)	36 ± 2	51 ± 1	44 ± 7	49 ± 7	62 ± 3
Sorghum (Sudan)	37 ± 4	22 ± 2	49 ± 5	82 ± 8	96 ± 1
Sorghum (Mali)	15 ± 2	14 ± 1	28 ± 7	56 ± 2	89 ± 2

These results also show that even if parasitic plant seed populations are able to germinate up to 100% (with GR24), they still can respond quite differently to the root exudates of host (or non-host) plants. We found similar differences in host specificity in several populations of *O. ramosa* collected from tomato, tobacco and rapeseed. Most of *O. ramosa* populations germinated to about 80% in a low (0.001 $mg.l^{-1}$) concentration of GR24 (maximum germination, in 0.1 $mg.l^{-1}$ GR24, 90-95%). However, the same tomato root exudates induced high germination of *O. ramosa* collected from tomato and tobacco fields but almost no germination of *O. ramosa* parasitizing rapeseed (data not shown). On the other hand, the *O. ramosa* collected from rapeseed germinated up to 90% after induction with the hairy-root exudates of *Arabidopsis*. It is obvious that there is some specificity in induction of germination by tomato (Solanaceae) or *Arabidopsis* (Brassicaceae) root exudates, depending on the host parasitized by the parent plant. However, Gurney et al. (2002) showed that host specificity is more complex and is also determined during later stages of the host–parasite interaction. In general, the seeds of *Orobanche* or *Striga* can germinate in the presence of several germination stimulants, but to a different extent. The germination in the presence of different host exudates gives the parasite an advantage of greater diversity of resources and ensures the survival of the parasite if the 'true' host is no longer present in the surrounding environment (Watling and Press 2001). The enormous amount of seeds produced by single plants of *Orobanche* or *Striga* spp. provides the best guarantee for the individual's contribution to future generations even if the most preferred host is not present (anymore).

CONCLUSION

This review summarizes what is known about the importance of the strigolactone germination stimulants in the interaction between host plants and the parasitic *Orobanche* and *Striga* spp. During preconditioning large changes in sensitivity of the parasitic plant seeds to the germination stimulants occur and there is an interesting analogy between these changes in sensitivity in parasitic plant seeds and the changes in sensitivity to other environmental and internal factors in their non-parasitic counterparts (Figure 3). These changes in dormancy may have ecological significance in restricting germination to the right period of the year. The selectivity of the response of parasitic plant seeds to specific germination stimulants may be one of the factors that determine host–parasite specificity. Finally, we have shown that the strigolactone germination stimulants are derived from the carotenoid biosynthetic pathway. This is a major breakthrough, although the primary function of these carotenoid-derived compounds remains unknown. Do these compounds have any function for the host or are they just breakdown products from the carotenoid pathway? It is of great interest to answer these questions, because the knowledge on the possible primary function of the germination stimulants will help to propose the most effective strategies to eliminate the parasite without a harmful impact on the host plant.

ACKNOWLEDGEMENTS

We thank Vicky Child for maize and *S. hermonthica* seeds as well as many helpful suggestions, the Maize Genetic COOP Stock Center for supplying seeds of maize mutants, Piet Arts of J.C. Robinson Seeds for Dent maize seeds, Bob Vasey for his kind help in supplying many different batches of host as well as parasite seeds and Danny Joel for supplying *O. crenata* seeds. This work was supported by the European Commission [the FP5 EU project Improved Striga Control in Maize and Sorghum (INCO-DEV, ICA4-CT-2000-30012) (to HJB) and the FP6 EU Project Grain Legumes (FOOD-CT-2004-506223) (to HJB and RM)]; the Netherlands Ministry of Agriculture, Nature and Food Quality in the form of an IAC fellowship (to RM) and the North-South programme (to HJB); the Netherlands Organisation for Scientific Research (NWO) (NATO visiting-scientist fellowships to RM); the Organisation for Economic Co-operation and Development OECD (a fellowship under the Co-operative Research Programme: Biological Resource Management for Sustainable Agriculture Systems [to RM]).

REFERENCES

Akhtouch, B., Munoz-Ruz, J., Melero-Vara, J., et al. 2002. Inheritance of resistance to race F of broomrape in sunflower lines of different origins. *Plant Breeding,* 121 (3), 266-268.

Akiyama, K., Matsuzaki, K. and Hayashi, H., 2005. Plant sesquiterpenes induce hyphal branching in arbuscular mycorrhizal fungi. *Nature,* 435 (7043), 824-827.

Aly, R., Goldwasser, Y., Eizenberg, H., et al. 2001. Broomrape (*Orobanche cumana*) control in sunflower (*Helianthus annuus*) with imazapic1. *Weed Technology,* 15 (2), 306-309.

Bais, H.P., Park, S.W., Weir, T.L., et al. 2004. How plants communicate using the underground information superhighway. *Trends in Plant Science,* 9 (1), 26-32.

Bouwmeester, H.J, Matusova, R., Zhongkui, S., et al. 2003. Secondary metabolite signalling in host-parasitic plant interactions. *Current Opinion in Plant Biology,* 6 (4), 358-364.

Butler, L.G., 1995. Chemical communication between the parasitic weed *Striga* and its crop host: a new dimention of allelochemistry. *In:* Inderjit, Dakshini, K.M.M. and Einhellig, F.A. eds. *Allelopathy: organisms, processes and applications*. American Chemical Society, Washington, 158-168. ACS Symposium Series no. 582.

Chang, M., Netzly, D.H., Butler, L.G., et al. 1986. Chemical regulation of distance: characterization of the first natural host germination stimulant for *Striga asiatica*. *Journal of the American Chemical Society,* 108 (24), 7858-7860.
Cook, C.E., Whichard, L.P., Wall, M.E., et al. 1972. Germination stimulants. 2. The structure of strigol-a potent seed germination stimulant for witchweed (*Striga lutea* Lour.). *Journal of American Chemical Society,* 94 (17), 6198-6199.
Dalla Vecchia, F., Barbato, R., La Rocca, N., et al. 2001. Responses to bleaching herbicides by leaf chloroplasts of maize plants grown at different temperatures. *Journal of Experimental Botany,* 52 (357), 811-820.
Degenhardt, J., Gershenzon, J., Baldwin, I.T., et al. 2003. Attracting friends to feast on foes: engineering terpene emission to make crop plants more attractive to herbivore enemies. *Current Opinion in Biotechnology,* 14 (2), 169-176.
Derkx, M.P.M. and Karssen, C.M., 1993a. Changing sensitivity to light and nitrate but not to gibberellins regulates seasonal dormancy patterns in *Sisymbrium Officinale* seeds. *Plant Cell and Environment,* 16 (5), 469-479.
Derkx, M.P.M. and Karssen, C.M., 1993b. Effects of light and temperature on seed dormancy and gibberellin-stimulated germination in *Arabidopsis thaliana*: studies with gibberellin-deficient and gibberellin-insensitive mutants. *Physiologia Plantarum,* 89 (2), 360-368.
Derkx, M.P.M. and Karssen, C.M., 1994. Are seasonal dormancy patterns in *Arabidopsis thaliana* regulated by changes in seed sensitivity to light, nitrate and gibberellin. *Annals of Botany,* 73 (2), 129-136.
Dicke, M. and Hilker, M., 2003. Induced plant defences: from molecular biology to evolutionary ecology. *Basic and Applied Ecology,* 4 (1), 3-14.
Dube, M.P. and Olivier, A., 2001. *Striga gesnerioides* and its host, cowpea: interaction and methods of control. *Canadian Journal of Botany,* 79 (10), 1225-1240.
Gibot-Leclerc, S., Corbineau, F., Salle, G., et al. 2004. Responsiveness of *Orobanche ramosa* L. seeds to GR 24 as related to temperature, oxygen availability and water potential during preconditioning and subsequent germination. *Plant Growth Regulation,* 43 (1), 63-71.
Goldwasser, Y., Hershenhorn, J., Plakhine, D., et al. 1999. Biochemical factors involved in vetch resistance to *Orobanche aegyptiaca*. *Physiological and Molecular Plant Pathology,* 54 (3/4), 87-96.
Gurney, A.L., Press, M.C. and Scholes, J.D., 2002. Can wild relatives of sorghum provide new sources of resistance or tolerance against *Striga* species? *Weed Research,* 42 (4), 317-324.
Hable, W.E., Oishi, K.K. and Schumaker, K.S., 1998. Viviparous-5 encodes phytoene desaturase, an enzyme essential for abscisic acid (ABA) accumulation and seed development in maize. *Molecular and General Genetics,* 257 (2), 167-176.
Hauck, C., Muller, S. and Schildknecht, H., 1992. A germination stimulant for parasitic flowering plants from *Sorghum bicolor*, a genuine host plant. *Journal of Plant Physiology,* 139 (4), 474-478.
Hemmerlin, A., Hoeffler, J.F., Meyer, O., et al. 2003. Cross-talk between the cytosolic mevalonate and the plastidial methylerythritol phosphate pathways in Tobacco Bright Yellow-2 cells. *Journal of Biological Chemistry,* 278 (29), 26666-26676.
Hilhorst, H.W.M., 1993. New aspects of seed dormancy. *In:* Côme, D. and Corbineau, F. eds. *Proceedings of the fourth international workshop on seeds: basic and applied aspects of seed biology, Angers, France, 20–24 July 1992*, Vol. 2. Université Pierre et Marie Curie, Paris, 571-579.
Hilhorst, H.W.M., Derkx, M.P.M. and Karssen, C.M., 1996. An integrating model for seed dormancy cycling: characterization of reversible sensitivity. *In:* Lang, G.A. ed. *Plant dormancy: physiology, biochemistry and molecular biology*. Cab International, Wallingford, 341-360.
Hilhorst, H.W.M. and Karssen, C.M., 1988. Dual effect of light on the gibberellin- and nitrate-stimulated seed germination of *Sisymbrium officinale* and *Arabidopsis thaliana*. *Plant Physiology,* 86 (2), 591-597.
Hilhorst, H.W.M., Smitt, A.J. and Karssen, C.M., 1986. Gibberelin-biosynthesis and -sensitivity mediated stimulation of seed germination of *Sisymbrium officinale* by red light and nitrate. *Physiologia Plantarum,* 67, 285-290.
Hirsch, A.M., Bauer, W.D., Bird, D.M., et al. 2003. Molecular signals and receptors: controlling rhizosphere interactions between plants and other organisms. *Ecology,* 84 (4), 858-868.
Hood, M.E., Condon, J.M., Timko, M.P., et al. 1998. Primary haustorial development of *Striga asiatica* on host and nonhost species. *Phytopathology,* 88 (1), 70-75.

Keyes, W.J., Taylor, J.V., Apkarian, R.P., et al. 2001. Dancing together: social controls in parasitic plant development. *Plant Physiology,* 127 (4), 1508-1512.

Labrousse, P., Arnaud, M.C., Serieys, H., et al. 2001. Several mechanisms are involved in resistance of *Helianthus* to *Orobanche cumana* Wallr. *Annals of Botany,* 88 (5), 859-868.

Lee, H.S. and Milborrow, B.V., 1997. Endogenous biosynthetic precursors of (+)-abscisic acid. IV. Biosynthesis of ABA from [H-2(n)]carotenoids by a cell-free system from avocado. *Australian Journal of Plant Physiology,* 24 (6), 715-726.

Li, Z.H., Matthews, P.D., Burr, B., et al. 1996. Cloning and characterization of a maize cDNA encoding phytoene desaturase, an enzyme of the carotenoid biosynthetic pathway. *Plant Molecular Biology,* 30 (2), 269-279.

Limpens, E. and Bisseling, T., 2003. Signaling in symbiosis. *Current Opinion in Plant Biology,* 6 (4), 343-350.

Losner-Goshen, D., Portnoy, V.H., Mayer, A.M., et al. 1998. Pectolytic activity by the haustorium of the parasitic plant *Orobanche* L. (*Orobanchaceae*) in host roots. *Annals of Botany,* 81 (2), 319-326.

Mangnus, E.M. and Zwanenburg, B., 1992. Tentative molecular mechanism for germination stimulation of *Striga and Orobanche* seeds by strigol and its synthetic analogs. *Journal of Agricultural and Food Chemistry,* 40 (6), 1066-1070.

Matúšová, R., Rani, K., Verstappen, F.W.A., et al. in press. The strigolactone germination stimulants of the plant-parasitic *Striga* and *Orobanche* spp. are derived from the carotenoid pathway. *Plant Physiology.*

Matúšová, R., Van Mourik, T. and Bouwmeester, H.J., 2004. Changes in the sensitivity of parasitic weed seeds to germination stimulants. *Seed Science Research,* 14 (4), 335-344.

Mayer, A.M. and Bar Nun, N., 1997. Germination of *Orobanche* seeds: some aspects of metabolism during preconditioning. *In:* Ellis, R.H., Black, M., Murdoch, A.J., et al. eds. *Basic and applied aspects of seed biology: proceedings of the fifth international workshop on seeds, Reading, 1995.* Kluwer Academic Publishers, Dordrecht, 633-639.

Mohamed, A., Ellicott, A., Housley, T.L., et al. 2003. Hypersensitive response to *Striga* infection in *Sorghum. Crop Science,* 43 (4), 1320-1324.

Muller, S., Hauck, C. and Schildknecht, H., 1992. Germination stimulants produced by *Vigna unguiculata* Walp cv. Saunders Upright. *Journal of Plant Growth Regulation,* 11 (2), 77-84.

Mwakaboko, A.S., 2003. *Synthesis and biological evaluation of new strigolactone analogues as germination stimulants for the seeds of the parasitic weeds Striga and Orobanche spp.* PhD Thesis University of Nijmegen. [http://webdoc.ubn.kun.nl/mono/m/mwakaboko_a/syntanbie.pdf].

Parker, C. and Riches, C.R., 1993. *Orobanche* species: the broomrape. *In:* Parker, C. and Riches, C.R. eds. *Parasitic weeds of the world: biology and control.* CAB International, Wallingford, 111-163.

Parry, A.D. and Horgan, R., 1992. Abscisic-acid biosynthesis in roots. 1. The identification of potential abscisic-acid precursors, and other carotenoids. *Planta,* 187 (2), 185-191.

Press, M.C., Scholes, J.D. and Riches, C.R., 2001. Current status and future prospects for management of parasitic weeds (*Striga* and *Orobanche*). *In:* Riches, C.R. ed. *The world's worst weeds: proceedings of an international symposium, Brighton, 2001.* British Crop Protection Council, Farnham, 71-88. British Crop Protection Council Symposium Proceedings no. 77.

Rasmann, S., Kollner, T.G., Degenhardt, J., et al. 2005. Recruitment of entomopathogenic nematodes by insect-damaged maize roots. *Nature,* 434 (7034), 732-737.

Reizelman, A. and Zwanenburg, B., 2002. An efficient enantioselective synthesis of strigolactones with a palladium-catalyzed asymmetric coupling as the key step. *European Journal of Organic Chemistry* (5), 810-814.

Rugutt, J.K. and Berner, D.K., 1998. Activity of extracts from nonhost legumes on the germination of *Striga hermonthica* seeds. *Phytomedicine,* 5 (4), 293-299.

Sato, D., Awad, A.A., Takeuchi, Y., et al. 2005. Confirmation and quantification of strigolactones, germination stimulants for root parasitic plants *Striga* and *Orobanche*, produced by cotton. *Bioscience Biotechnology and Biochemistry,* 69 (1), 98-102.

Schwartz, S.H., Tan, B.C., Gage, D.A., et al. 1997. Specific oxidative cleavage of carotenoids by VP14 of maize. *Science,* 276 (5320), 1872-1874.

Serghini, K., Pérez de Luque, A., Castejón-Muñoz, M., et al. 2001. Sunflower (*Helianthus annuus* L.) response to broomrape (*Orobanche cernua* Loefl.) parasitism: induced synthesis and excretion of 7-hydroxylated simple coumarins. *Journal of Experimental Botany,* 52 (364), 2227-2234.

Siame, B.A., Weerasuriya, Y., Wood, K., et al. 1993. Isolation of strigol, a germination stimulant for *Striga asiatica*, from host plants. *Journal of Agricultural and Food Chemistry,* 41 (9), 1486-1491.

Tan, B.C., Schwartz, S.H., Zeevaart, J.A.D., et al. 1997. Genetic control of abscisic acid biosynthesis in maize. *Proceedings of the National Academy of Sciences of the United States of America,* 94 (22), 12235-12240.

Van Hezewijk, M.J., Van Beem, A.P., Verkleij, J.A.C., et al. 1993. Germination of *Orobanche crenata* seeds, as influenced by conditioning temperature and period. *Canadian Journal of Botany,* 71 (6), 786-792.

Vleeshouwers, L.M., Bouwmeester, H.J. and Karssen, C.M., 1995. Redefining seed dormancy: an attempt to integrate physiology and ecology. *Journal of Ecology,* 83 (6), 1031-1037.

Watling, J.R. and Press, M.C., 2001. Impacts of infection by parasitic angiosperms on host photosynthesis. *Plant Biology,* 3 (3), 244-250.

Wigchert, S.C.M. and Zwanenburg, B., 1999. A critical account on the inception of *Striga* seed germination. *Journal of Agricultural and Food Chemistry,* 47 (4), 1320-1325.

Yoder, J.I., 2001. Host-plant recognition by parasitic *Scrophulariaceae. Current Opinion in Plant Biology,* 4 (4), 359-365.

Yokota, T., Sakai, H., Okuno, K., et al. 1998. Alectrol and Orobanchol, germination stimulants for *Orobanche minor*, from its host red clover. *Phytochemistry,* 49 (7), 1967-1973.

Yoneyama, K., Takeuchi, Y., Sato, D., et al. 2004. Determination and quantification of strigolactones. *In: Proceedings of the 8th international parasitic weeds symposium, Durban (South Africa), June 24-25, 2004.* The International Parasitic Plant Society.

CHAPTER 5

CHEMICAL SIGNALLING BETWEEN PLANTS

Mechanistic similarities between phytotoxic allelopathy and host recognition by parasitic plants

ALEXEY TOMILOV, NATALYA TOMILOVA, DONG HYUN SHIN, DENNEAL JAMISON, MANUEL TORRES, RUSSELL REAGAN, HEATHER MCGRAY, TIZITA HORNING, RUTH TRUONG, AJ NAVA, ADRIAN NAVA AND JOHN I. YODER

Corresponding author: John I. Yoder, Department of Plant Sciences, University of California–Davis, Davis, CA 95616, USA. E-mail: jiyoder@ucdavis.edu

Abstract. Parasitic plants in the *Orobanchaceae* use chemicals released from host-plant roots to direct developmental processes crucial to their heterotrophic lifestyle. An illustrative example is the development of haustoria; parasite root organs that function in host attachment and penetration, and in the establishment of a physiological conduit through which host resources are robbed. The facultative parasite *Triphysaria* develops haustoria only in the presence of host roots or host root factors. An *in vitro* assay was used to identify several phenolic derivatives that induce haustorium formation; the activity of multiple signalling molecules is consistent with a redundancy of active molecules in the rhizosphere triggering haustorium development. Haustorium-inducing factors are structurally related to phytotoxic allelochemicals released by some plants to inhibit the growth of neighbouring plants. We used genomic approaches to demonstrate that similar genetic pathways are up-regulated in parasitic roots upon contact with host plants as are regulated in response to allelochemical exposure. A parasite quinone oxidoreductase was identified that has properties suggesting that it functions in both allelochemical detoxification and haustorium signal transduction. These and other mechanistic similarities between allelopathic toxicity and haustorium signal transduction support the hypothesis that parasitic plants have recruited allelotoxin defence mechanisms for host-plant recognition.
Keywords: parasitic plants; allelopathy; plant–plant communication; haustorium development

INTRODUCTION

Parasitic angiosperms live in intimate associations with their plant hosts and by their very definition fulfil at least some of their nutritional requirements by directly invading other plants to rob them of water and nutrients (Kuijt 1969). In some species host-plant identification and invasion is orchestrated through chemical signalling between the host and parasite. Most notably, parasitic species of

M. Dicke and W. Takken (eds.), Chemical Ecology: From Gene to Ecosystem, 55-69.

Orobanchaceae use molecules made by the host root to trigger various developmental programmes including seed germination, host attachment and invasion, and the establishment of physiological conduits through which nutrients are transferred from host to parasite (Press and Graves 1995). The effects of plant parasitism can be devastating for the host plants and some of the world's worst agricultural pests are parasitic weeds (Parker and Riches 1993; Matúšová and Bouwmeester in press).

Almost thirty years ago Peter Atsatt drew parallels between insect herbivory and plant parasitism and suggested that parasitic plants, like specialist insect herbivores, may recruit plant defence molecules as 'feeding cues' (Atsatt 1977). This insightful analogy was proposed before most of the molecules used by parasitic plants for host recognition were identified. As seen in Figure 1, many of the molecules used by parasitic plants for host identification are structurally similar to phytotoxins produced by allelopathic plants to inhibit the growth of neighbouring plants (Conger 1999). We will show that not only are phytotoxic allelochemicals similar to host recognition cues, but in some cases the same molecules have both activities. We will also show that many of the genes activated in parasite roots after contact with host roots are similarly activated by exposure to allelotoxins. At least two parasite genes activated by host root contact encode quinone oxidoreductases that are known to function in xenobiotic detoxification in other biological systems. Biochemical and transcriptional experiments suggest that one of these may also function in host signal perception and transduction. While the molecular mechanisms of host identification by parasitic plants have yet to be fully elucidated, the current evidence suggests underlying similarities between host plant recognition and defence against allelopathic phytotoxins. The collective conclusions of these studies support Atsatt's hypothesis that parasitic plants have adapted host defence molecules as recognition cues.

PHYTOTOXIC ALLELOPATHY

For many years it was accepted that spatial patterning of plants in natural populations is established to a large extent by inherent properties of the plants themselves. There was, however, considerable debate about the role of chemical factors in establishing localized communities (Muller et al. 1964). While there were numerous publications of phytotoxic molecules being produced by plants, a phenomenon generally termed allelopathy, the ecological and/or agronomic effects of these molecules in field settings remained questionable (Conger 1999; Williamson 1990). Recently it was shown that the toxic flavonoid catechin is secreted into the soil by *Centaurea maculoso*, an exotic invasive weed of North America (Bais et al. 2002). Native grasses in North America are more susceptible to catechin than are their European relatives. Also, catechin concentrations are higher in North-American grasslands invaded by *C. maculoso* than in European grasslands where *C. maculoso* is native. The conclusion reached by these studies is that secretion of phytotoxic catechin contributes to the invasive success of this pernicious weed and established that allelotoxins exchanged between plants are of ecological significance (Bais et al. 2003; 2002). However, our ability to exploit allelopathic phytotoxins in agricultural settings remains limited by a general lack of knowledge about mechanisms underlying plant–plant interactions.

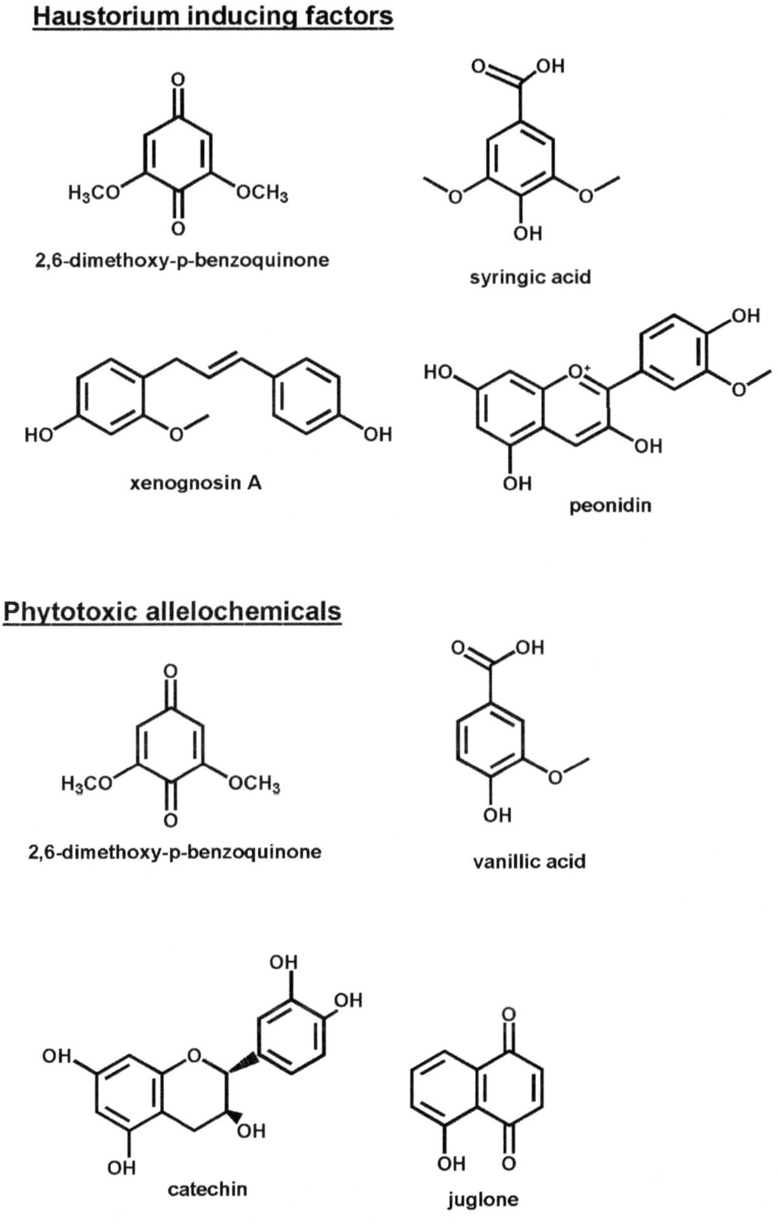

Figure 1. *Common haustorial inducing and phytotoxic allelochemicals*

It has been known for centuries that walnut trees poison the soil for underlying vegetation (Gries 1942). The allelochemical responsible for walnut toxicity is juglone (5,hydroxyl-1,4-naphthoquinone), a highly toxic quinone frequently used in pharmacological studies (Gries 1942; Inbaraj and Chignell 2004; Kamei et al. 1998) (Figure 1). We assayed the effects of juglone on *Arabidopsis* seed germination and root growth. Germination was assayed by plating the seeds directly into media containing various concentrations of juglone; root growth was measured by germinating the seeds in non-selective media and then transplanting the seedlings into juglone-containing media (Figure 2, top). As seen from the bars in Figure 3, there was a significant reduction (T test, $P \leq 0.05$) in both germination and root growth rates in juglone concentrations greater than 40 μM. The concentrations of juglone required for ½ maximal germination or growth were similar, suggesting that phytotoxicity is associated with a common metabolic pathway shared by germination and root growth processes.

Quinones and phenolics are among the most commonly described classes of allelopathic phytotoxins (Inderjit 1996) (Figure 1). Quinones are oxidized phenols, and phenols are reduced quinones, and electrical transformations between these states account for much of their biological significance (Harborne 1989). Because quinones are widely used in medicine as anticancer agents, antibiotics and antimalarial drugs, the mechanisms of quinone cytotoxicity are well known (O'Brien 1991). Most significant are those mechanisms associated with free-radical formation during quinone reduction. Single electron reductions catalysed by enzymes such as quinone oxidoreductase or xanthine oxidase produce highly reactive semiquinone intermediates that directly bind to and inactivate nucleic acids, proteins, lipids and carbohydrates (Testa 1995). Semiquinone radicals also react with molecular oxygen leading to the generation of superoxide anions and hydroxyl radicals. These highly toxic radicals inactivate enzymes, break DNA strands, and cause membrane-lipid peroxidation. These molecules also play an integral role in the cytotoxicity associated with the hypersensitivity response of plants against microbial pathogens (Hammond-Kosack and Jones 1996).

There are good reasons to believe that juglone phytotoxicity results from similar mechanisms. Juglone is not synthesized by walnut trees, which rather synthesize the non-toxic reduced form 1,4,5-trihydroxynaphthalene (hydrojuglone) (Lee and Campbell 1969). Hydrojuglone is abundantly produced by roots, leaves and nuts and becomes oxidized to toxic juglone upon exposure to air or oxidizing agents from other organisms, including roots of other plants (Gries 1942). Free radicals formed during redox cycling between juglone and hydrojuglone have been identified in human and mouse cells and intact *Caenorhabditis elegans* (Chignell and Sik 2003; Noda et al. 1997; De Castro et al. 2004). While the cytotoxicity mechanisms of this and other phenolic allelotoxins have not been specifically elucidated, it is reasonable to propose that toxicity is to a large extent associated with free radicals produced during redox cycling.

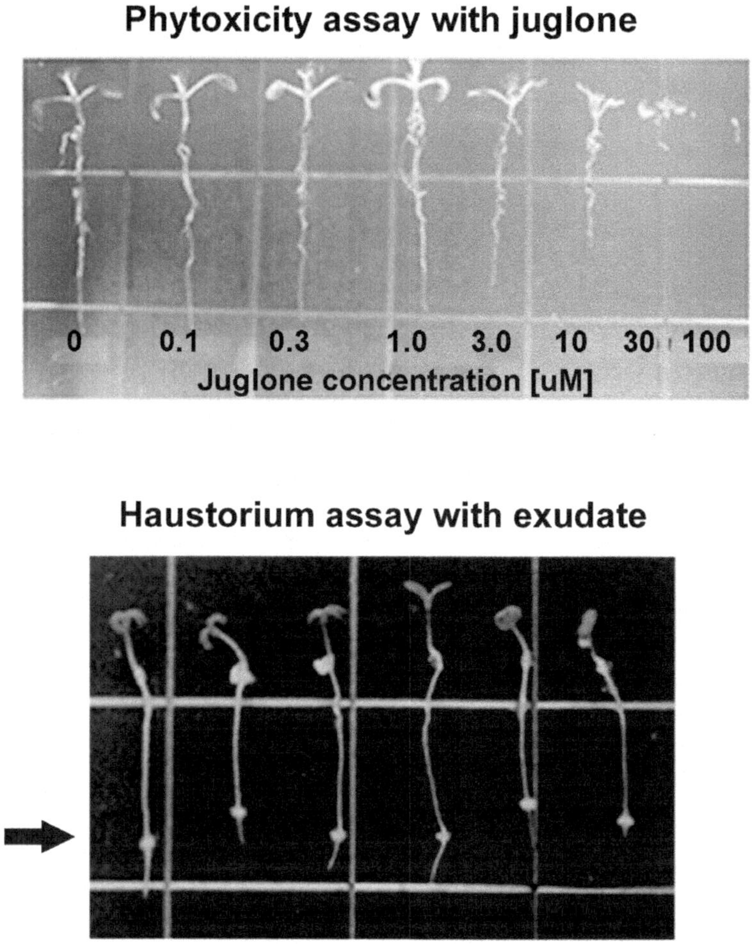

Figure 2. *Phytotoxicity and haustorium induction assays. Top photo: Aseptic Arabidopsis seedlings were placed in media containing juglone at the concentrations indicated. After nine days the seedlings were removed, spread along the surface of an agar plate and photographed. Bottom photo: Aseptic Triphysaria seedlings were germinated in agar, exposed to rice exudates, and photographed thirty hours later. The arrow approximately marks the single haustorium formed on every Triphysaria root*

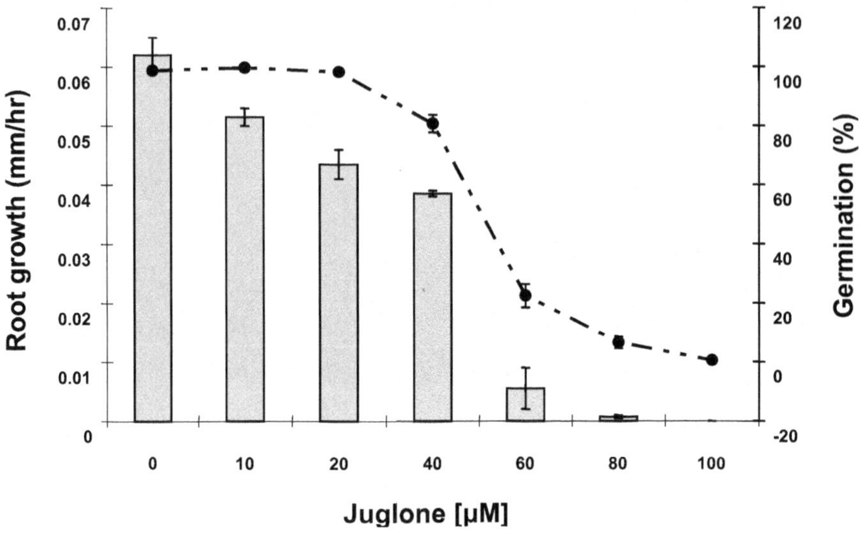

Figure 3. *Toxicity of juglone on Arabidopsis germination and growth. The bars indicate Arabidopsis root growth rates in different concentrations of juglone and are referenced by the primary axis. About 35 roots were measured in each of two experiments for each data point graphed. The dashed line shows the percent germination at the same juglone concentrations. The results are the averages of two experiments with about 300 seeds each. The error bars indicate the minimum and maximum values obtained*

HOST RECOGNITION AND HAUSTORIUM DEVELOPMENT IN THE PARASITIC PLANT *TRIPHYSARIA*

Over three thousand angiosperm species are parasitic and able to invade host plants to obtain nutrients (Nickrent 2005). Parasitic plants encompass a wide range of growth habits ranging from mistletoes that grow on the tops of conifers to root parasites, like *Striga*, that live a significant portion of their lives underground. Perhaps the most bizarre habit is displayed by *Rafflesia*, a rootless, stemless plant comprised of little more than the world's biggest flower (Brown 1822). The single morphological feature that all parasitic plants have in common is their ability to produce a haustorium, a structure able to invade host plant tissues and act as the physiological bridge through which host resources are translocated into the parasite (Kuijt 1969).

At least one family of parasitic angiosperms, the *Orobanchaceae*, develops haustoria in response to molecules secreted by host plant roots. This family is comprised of about thirty species of root parasites that rely on host resources to varying degrees. Representative of the obligate parasites that must attach to host plants within days of germination are the agriculturally devastating weeds *Striga* and *Orobanche* (Parker and Riches 1993). As described elsewhere in this volume, these plants have evolved host detection systems to identify host roots prior to committing to germination (Matúšová and Bouwmeester in press). Other *Orobanchaceae* are facultative parasites that do not require host germination factors and can mature without attaching to a host. Facultative parasites do, however, require host factors to initiate the switch from autotrophic to heterotrophic growth.

Triphysaria is a facultative parasite that grows as a common springtime annual throughout the pacific coast of North America. *Triphysaria* is a small genus with five inter-hybridizing species, four of which are outcrossing and one autogenous (Yoder 1998). We are using *Triphysaria* to study the genetic factors that govern plant parasitism because unlike *Striga*, *Triphysaria* can be grown in the US without quarantine restriction or environmental concerns. This allows us to easily collect large numbers of seeds that represent a wide range of genetic variants. *Triphysaria* has a broad host range that includes at least 27 families of angiosperms ranging from *Arabidopsis* to maize (Thurman 1966; Goldwasser et al. 2002). Intriguingly, the only plant species apparently not infected by *Triphysaria* are other *Triphysaria* (Yoder 1997). The mechanism of vegetative self-recognition in *Triphysaria* is not currently known but is an active area of investigation because of its potential application for engineering host resistance against parasitic weeds.

Haustorium development in *Triphysaria* roots can be monitored *in vitro* by applying host roots, root exudates or purified root factors to aseptic *Triphysaria* seedlings (Jamison and Yoder 2001). In brief, *Triphysaria* seeds are surface-sterilized and germinated in agar plates at 16°C. One to two weeks after germination, aseptic seedlings are transferred to square Petri dishes containing nutrient agar, and incubated at 20°C at a near vertical angle so that the *Triphysaria* roots grow down along the agar surface. After additional one or two weeks of growth, host root exudates or aqueous solutions of purified haustoria-inducing factors (HIFs) are spread across the roots. The first morphological response to HIF exposure is an almost immediate cessation of root elongation (Baird and Riopel 1984). Within about five hours haustorial hairs begin to proliferate in a zone just behind the root tip. Concomitantly, cortical cells underlying the haustorial hairs begin to expand and by twelve hours a hairy, swollen knob appears distal to the root tip. In the presence of a host, haustorial hairs will attach themselves firmly to the host root and the haustorium will penetrate via a combination of enzymatic activity and physical pressure (Losner-Goshen et al. 1998). In the absence of a host root the swelling and hair proliferation continue for about 24 hours at which time the *Triphysaria* root reverts to its normal growth programme. Haustorium development is highly synchronous, and when several *Triphysaria* are treated together haustoria are observed at defined locations distal to the tip (Figure 2). Photographs of haustoria and a time-lapse animation of haustorium development can be seen at http://www.plantsciences.ucdavis.edu/yoder/lab/.

Using the *in vitro* assay we identified several phenolic derivatives that trigger haustorium formation when applied to *Triphysaria* roots including simple phenolics, flavonoids and quinones (Figure 1) (Albrecht et al. 1999). Similar molecules were previously identified as HIFs for *Striga* and *Agalinis* (Riopel and Timko 1995; Smith et al. 1996). Many of these molecules are commonly found in the rhizosphere and play signalling roles in the attraction and/or repulsion of microbial populations (Siqueira et al. 1991). The triggering of haustorium development by multiple phytochemical signals suggests that there is a redundancy in HIFs functioning in the rhizosphere. This hypothesis is supported by our observation that inbred lines of *Triphysaria* selected for the inability to form haustoria when exposed to a specific HIF still form haustoria when exposed to complex host root exudates (Jamison and Yoder 2001).

Two general hypotheses can be proposed for the ability of *Triphysaria* to form haustoria in response to several different molecules. One hypothesis is that there are several specific receptors, each recognizing a different HIF, that trigger haustorium development. Alternatively there may be a single receptor that recognizes multiple inducing molecules. Because HIF receptors have not yet been isolated we cannot rule out either mechanism. However an informative set of experiments conducted by David Lynn and co-workers suggests a model for activation of a single receptor by multiple phenolics. This group assayed a number of natural and synthetic quinones for their ability to induce haustoria in *Striga* (Smith et al. 1996). Active haustorial-inducing quinones had similar redox potentials while inactive molecules generally fell outside the redox window. Lynn's group then designed spin trap molecules that acted as haustorium development inhibitors (Zeng et al. 1996). This work led them to suggest that haustorium signalling involves a redox-regulated signalling mechanism that is triggered by cycling between the reduced and oxidized states of the HIF. There is considerable precedent for redox regulation of development and many biological processes are under redox control, including DNA replication, transcription, translation, hormone reception, phototropism and defence responses (Huala et al. 1997; Allen 1993).

Redox cycling of quinones is catalysed by quinone oxidoreductases and, as will be discussed later, we have studied two *Triphysaria* quinone oxidoreductases that are active during haustorium initiation. The role of quinone oxidoreductases in haustorium signalling was examined using pharmacological inhibitors (Matvienko et al. 2001b). Dicumarol and Cibacron blue are specific inhibitors of quinone oxidoreductases, and these inhibit haustorium formation when applied to *Triphysaria* roots prior to host root factors. Root growth measurements taken before and after inhibitor exposure showed that the inhibitors did not affect overall root health. These experiments support the model that enzymatically catalysed quinone oxidoreduction is a component of haustorium signalling (Matvienko et al. 2001b).

The current model for haustorium initiation predicts that semiquinone intermediates formed from the action of quinone oxidoreductase initiate haustorium signal transduction through a redox signalling pathway. This model has obvious parallels to the mechanisms of allelopathic quinone toxicity since both are dependent upon the generation of free-radical intermediates. The important roles of redox transformations in subterranean interactions between plants and other rhizosphere organisms have been previously highlighted (Appel 1993).

HAUSTORIUM INDUCING FACTORS CAN BE ALLELOTOXINS

Host root factors can be both phytotoxic and organogenic. We collected root exudates from hydroponically grown rice, bound small molecules to the non-ionic absorbent Bio-Beads SM2, and eluted them with methanol. After the methanol was evaporated, the dried exudate material was dissolved in water and diluted to concentrations either more or less concentrated than the original exudate. The diluted exudates were then applied to roots of *Triphysaria* seedlings as described for the haustorium bioassays. Phytotoxicity was estimated after three days by visually examining the roots and noting the degree of browning. Additionally, cell viability was assayed by staining the roots with fluorescein diacetate (FDA) and monitoring the loss of fluorescence as the dye leaked from dead cells (Bais et al. 2003). Figure 4 summarizes the results (Shin and Yoder in prep.). Haustorium formation was maximal with about 90% of the roots forming haustoria at original, undiluted exudate concentration (1X). As exudate concentrations increased, haustorium formation decreased with a concomitant increase in cytotoxicity by both direct visualization and loss of FDA staining. At exudate concentrations six times that of the original, *Triphysaria* roots did not develop haustoria and were beginning to turn brown (Figure 4).

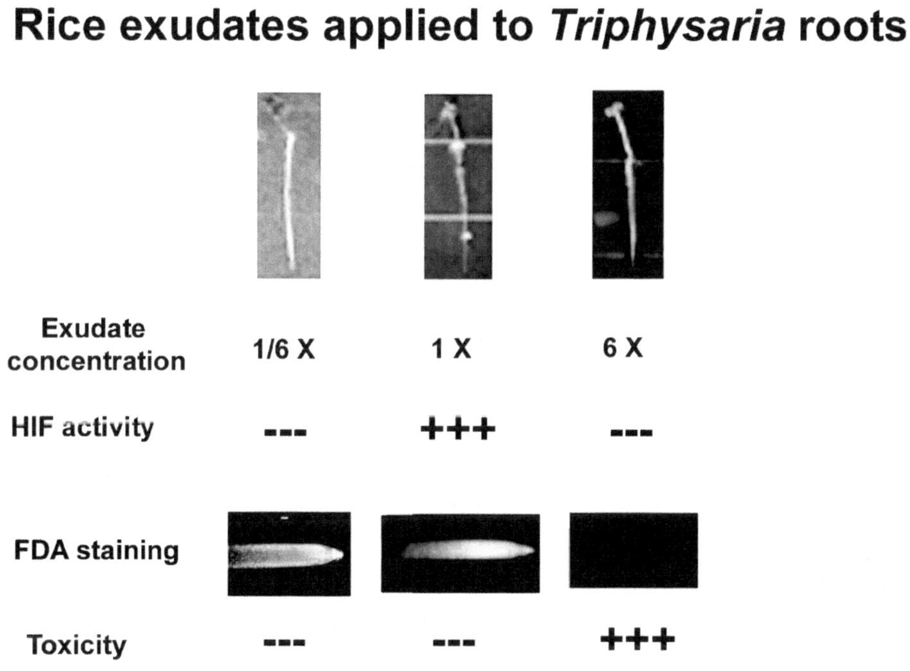

Figure 4. Rice root exudates have both HIF and phytotoxicity activities. Triphysaria seedlings were treated with different concentrations of rice root exudates and scored for haustorium formation and toxicity using FDA staining and root browning as indicators. The three photos at the top of the figure are representative of seedlings treated with 1/6X, 1X and 6X concentrations with 1/6X, 1X and 6X concentrations of exudate

Purified haustorial inducing molecules are also phytotoxic at high concentrations. The frequently referenced HIF isolated from sorghum roots, 2,6 dimethoxybenzoquinone (DMBQ), is an illustrative example (Chang and Lynn 1986). DMBQ was originally characterized as a mammalian cell cytotoxin and later as a microbial antibiotic and DNA mutagen (Nishina et al. 1991; Brambilla et al. 1988; Jones et al. 1981). We showed that while DMBQ is an active inducer of *Triphysaria* haustoria at concentrations between one and thirty μM, at concentrations one hundred μM or higher it is phytotoxic and *Triphysaria* roots turn brown and die (Jamison and Yoder 2001). In conclusion, both complex root exudates and purified factors can have either haustorium-inducing or phytotoxic activities depending on their concentrations.

TRIPHYSARIA GENES REGULATED BY HOST CONTACT FUNCTION IN ALLELOCHEMICAL DETOXIFICATION

A second factor linking host-parasite recognition and allelotoxin defence is the overlap in transcripts differentially regulated in each system. This was discovered by analysing the sequences of cDNA libraries enriched for transcripts regulated in *Triphysaria* roots after contact with host roots or DMBQ (Tomilov, Tomilova and Yoder in prep.Matvienko et al. 2001a). In brief, host contact was realized by laying the roots of *Arabidopsis* seedlings across those of *Triphysaria* growing along the surface of agar plates. The *Arabidopsis* seedlings were removed at various times ranging from immediately after contact to up to five hours later. These times correlated with early haustorium development prior to host-root penetration. *Triphysaria* roots were then dissected, frozen in liquid nitrogen and subjected to mRNA isolation. PCR-based suppression subtractive hybridization (SSH) was used to prepare two cDNA libraries, one enriched for transcripts up-regulated (host forward, 'HF') and one enriched for transcripts down-regulated (host reverse, 'HR') by contact with *Arabidopsis* roots (Diatchenko et al. 1996). Approximately 3000 inserts of each library were sequenced and assembled into contigs representing over 1000 distinct transcripts in each class.

BLASTN analyses showed that approximately 80% of the cDNAs were specific to one or the other library. We assigned a tentative function to each cDNA by virtually translating the assembled transcripts and comparing the predicted proteins to those catalogued in the *Arabidopsis* protein database (ATH1.pep_cm_20040228) using BLASTX (Rhee et al. 2003). The corresponding GO annotations for each of the best hits was obtained through the Gene Ontology (GO) function at TAIR (TAIR 2005). GO annotations provide a uniform vocabulary to describe the roles of genes and gene products in all organisms and allowed us to categorize the putative functions of each translation product into one of nine general biological processes (Ashbuner et al. 2000). The number of transcripts in each category for either the HF or HR libraries allowed us to determine which biological functions were over- or under-represented in each library. Three classes of transcripts were significantly ($p < 0.01$) more abundant in the HF than HR libraries; those involved in stress responses, electron transport or cellular transport (Table 1). As previously observed, many of these transcripts function in xenobiotic detoxification and/or protection from reactive oxygen species (Matvienko et al. 2001a).

Table 1. *Representation of biological functions in different SSH libraries*

	HF[1]	HR[2]	Chi^2	P
Total # annotated transcripts	**702**	**910**		
DNA or RNA metabolism	28	44	0.67	NS
cell organization and biogenesis	44	48	0.73	NS
electron transport or energy pathways	93	64	17.41	$p < 0.001$
protein metabolism	174	220	0.08	NS
signal transduction	28	35	0.02	NS
transcription	43	48	0.54	NS
transport	157	153	7.86	$p < 0.01$
response to abiotic or biotic stimulus	58	47	6.24	NS
response to stress	52	34	10.58	$p < 0.001$

[1] Host forward subtracted library
[2] Host reverse subtracted library
Chi^2 and P show significance values for the functional category being differentially represented in either the HF or HR libraries.

We are interested in genes predicted to function in allelochemical oxidoreduction because of their hypothesized roles in haustorium initiation and allelopathic phytotoxicity. Two distinct quinone oxidoreductases were selected from the SSH libraries and studied in detail (Wrobel et al. 2002; Matvienko et al. 2001b). TvQR2 encodes a 205 aa protein with significant homology to a quinone oxidoreductase in the wood-rotting fungus *Phanerochaete chrysosporium*. The *P. chrysosporium* quinone oxidoreductase functions to protect the fungus from the variety of toxic electrophiles produced during lignin degradation (Brock and Gold 1996). These enzymes are related to the carcinogen detoxification enzyme DT-diaphorase that reduces quinones to non-toxic hydroquinones by catalysing two-step hydride transfers from NAD(P)H to enzyme-bound FMN (or FAD), and then from FMNH2 (or FADH2) to the quinone. These detoxifying quinone reductases thereby reduce quinones to hydroquinones in a single-step reaction that avoids radical intermediates (Li et al. 1995).

TvQR1 encodes a 329 aa protein related to a family of NAD(P)H-dependent quinone oxidoreductases that produce semiquinone radicals through univalent quinone reductions. These enzymes catalyse the reduction of several natural quinines and have been identified in plants, animals and microbes (Babiychuk et al. 1995; Thorn et al. 1995). Electron paramagnetic resonance spectroscopy indicates that these enzymes catalyse single electron reductions that yield unstable semiquinone intermediates (Rao et al. 1992). The activated semiquinones are then readily detoxified by modifications with various chemical groups (Testa 1995).

We expressed and purified the TvQR1 protein from *E. coli* and the TvQR2 protein from *Pichia pastoris*. We spectrophotometrically monitored the reduction of quinone substrates and the oxidation of NADH to show that these enzymes catalyse NAD(P)H-dependent reductions of DMBQ, juglone and other allelopathic quinones (Wrobel et al. 2002, Petit and Yoder unpubl.). The biochemical analyses confirmed the homology predictions that these enzymes function in allelochemical detoxification.

Northern analyses showed that the steady-state transcript levels of TvQR1 and TvQR2 increased within 30 minutes of treatment with DMBQ, 2,6-dimethylbenzoquinone, menadione and, most strongly, juglone (Matvienko et al. 2001b). Steady-state levels reached a maximum 8-12 hours after treatment and returned to non-induced levels by 24 hours post-treatment, precisely corresponding to the times of haustorium ontogeny. The protein-synthesis inhibitor cycloheximide prevented haustorium development when applied to *Triphysaria* roots prior to host factors indicating that *de novo* protein synthesis is required for haustorium development. However, cycloheximide did not block transcriptional induction of TvQR1 or TvQR2 indicating that their transcriptional regulation is a rapid, primary response to both HIFs and allelochemical cytotoxins (Matvienko et al. 2001a).

Similar Northerns were performed after exposing roots of three non-parasitic *Scrophulariaceae*, *Lindenbergia muraria*, the closest non-parasite to the parasitic clade of *Scrophulariaceae* (DePamphilis et al. 1997), *Mimulus aurantiacus* and *Antirrhinum majus*, to DMBQ. TvQR2 homologues were induced in all species. In contrast, TvQR1 was only up-regulated in roots of parasitic species. Moreover, TvQR1 was up-regulated in response to DMBQ application in inbred lines of *T.pusilla* that formed haustoria but not in those selected to be non-responsive to DMBQ (Jamison 2003). The correlation between the up-regulation of TvQR1 and haustorium development holds for intraspecific as well as intergenic comparisons.

The correlation of TvQR1 transcript regulation with haustorium development together with its biochemical function suggests that this enzyme may play a role in haustorium formation. We hypothesize that semiquinone radicals produced by univalent quinone reductions catalysed by TvQR1 initiate the signal transduction pathway leading to haustorium development. Alternatively, semiquinone radicals and associated reactive oxygen intermediates may take a more direct role in early haustorium development. For example, cortical cell swelling and epidermal hair elongation may directly reflect the action of reactive radicals produced by over-expression of TvQR1. In either case, the induction of a univalent reducing quinone oxidoreductase by haustorium-inducing factors may be a critical developmental step that distinguishes parasitic plants from non-parasitic autotrophs. The development of a *Triphysaria* transient transformation system will allow us to test these hypotheses using inhibitory RNAs (Tomilov et al. 2004, Tomilov, Tomilova and Yoder in prep.).

CONCLUSIONS

Allelopathic plants release phytotoxic molecules into the soil as a means of limiting the growth of other plants. These can be thought of as molecules that defend allelopathic plants against neighbouring plants that compete for limiting resources. The phytotoxicity of these molecules results primarily from reactive oxygen species generated during redox cycling between reduced and oxidized states of the allelochemical. Plants and other organisms encode enzymes that detoxify reactive oxygen species; these protein families originated early in evolutionary history in defence against damage associated with aerobic environments (Testa 1995). Parasitic plants seem to have recruited some of the enzymes that function in xenobiotic detoxification for use in host root identification. Conclusive evidence that the parasite host recognition system is derived from an allelochemical detoxification system awaits gene-silencing experiments in transgenic parasites. But in any case, host defence and host recognition are clearly associated in parasitic plants and Atsatt's analogies between parasitic plants and herbivorous insects have to date withstood molecular investigations.

ACKNOWLEDGEMENTS

This work was supported by NSF grant #0236545. T. Horning, R. Truong, A.J. Nava and A. Nava were supported by REU supplements. D.H. Shin was supported by the L.G. Yonam Foundation.

REFERENCES

Albrecht, H., Yoder, J.I. and Phillips, D.A., 1999. Flavonoids promote haustoria formation in the root parasite *Triphysaria versicolor*. *Plant Physiology,* 119 (2), 585-591.

Allen, J.F., 1993. Redox control of transcription: sensors, response regulators, activators and repressors. *FEBS Letters,* 332 (3), 203-207.

Appel, H.M., 1993. Phenolics in ecological interactions: the importance of oxidation. *Journal of Chemical Ecology,* 19 (7), 1521-1552.

Ashbuner, M., Ball, C.A., Blake, J.A., et al. 2000. Gene ontology: tool for the unification of biology. *Nature Genetics,* 25 (1), 25-29.

Atsatt, P.R., 1977. The insect herbivore as a predictive model in parasitic seed plant biology. *American Naturalist,* 111 (979), 579-586.

Babiychuk, E., Kushnir, S., Bellesboix, E., et al. 1995. *Arabidopsis thaliana* NADPH oxidoreductase homologs confer tolerance of yeasts toward the thiol-oxidizing drug diamide. *Journal of Biological Chemistry,* 270 (44), 26224-26231.

Baird, W.V. and Riopel, J.L., 1984. Experimental studies of haustorium initiation and early development in *Agalinis purpurea* (L.) Raf. (Scrophulariaceae). *American Journal of Botany,* 71 (6), 803-814.

Bais, H.P., Vepachedu, R., Gilroy, S., et al. 2003. Allelopathy and exotic plant invasion: from molecules and genes to species interactions. *Science,* 301 (5638), 1377-1380.

Bais, H.P., Walker, T.S., Stermitz, F.R., et al. 2002. Enantiomeric-dependent phytotoxic and antimicrobial activity of (+or-)-catechin: a rhizosecreted racemic mixture from spotted knapweed. *Plant Physiology,* 128 (4), 1173-1179.

Brambilla, G., Robbiano, L., Cajelli, E., et al. 1988. Cytotoxic DNA-damaging and mutagenic properties of 2,6-dimethoxy-1,4-benzoquinone, formed by dimethophrine-nitrite interaction. *Journal of Pharmacology and Experimental Therapeutics,* 244 (3), 1011-1015.

Brock, B.J. and Gold, M.H., 1996. 1,4-benzoquinone reductase from the basidiomycete *Phanerochaete chrysosporium*: spectral and kinetic analysis. *Archives of Biochemistry and Biophysics,* 331 (1), 31-40.

Brown, R., 1822. An account of a new genus of plants named *Rafflesia*. *Transactions of the Linnean Society London,* 13, 201-234.

Chang, M. and Lynn, D.G., 1986. The haustorium and the chemistry of host recognition in parasitic angiosperms. *Journal of Chemical Ecology,* 12 (2), 561-579.

Chignell, C.F. and Sik, R.H., 2003. A photochemical study of cells loaded with 2',7'-dichlorofluorescin: implications for the detection of reactive oxygen species generated during UVA irradiation. *Free Radical Biology and Medicine,* 34 (8), 1029-1034.

Conger, B.V. (ed.) 1999. *Special issue on allelopathy*. Critical Reviews in Plant Sciences, 18 (5).

De Castro, E., Hegi de Castro, S. and Johnson, T.E., 2004. Isolation of long-lived mutants in *Caenorhabditis elegans* using selection for resistance to juglone. *Free Radical Biology and Medicine,* 37 (2), 139-145.

DePamphilis, C.W., Young, N.D. and Wolfe, A.D., 1997. Evolution of plastid gene rps2 in a lineage of hemiparasitic and holoparasitic plants: many losses of photosynthesis and complex patterns of rate variation. *Proceedings of the National Academy of Sciences of the United States of America,* 94 (14), 7367-7372.

Diatchenko, L., Lau, Y.F., Campbell, A.P., et al. 1996. Suppression subtractive hybridization: a method for generating differentially regulated or tissue-specific cDNA probes and libraries. *Proceedings of the National Academy of Sciences of the United States of America,* 93 (12), 6025-6030.

Goldwasser, Y., Westwood, J.H. and Yoder, J.I., 2002. The use of *Arabidopsis* to study interactions between parasitic angiosperms and their plant hosts. *In:* Somerville, C.R. and Meyerowitz, E. eds. *The Arabidopsis Book.* American Society of Plant Biologists, Rockville, MD. [http://www.bioone.org/pdfserv/i1543-8120-023-01-0001.pdf]

Gries, G., 1942. Juglone, the active agent in walnut toxicity. *National Nut Growers Association Annual Report,* 33, 52-55.

Hammond-Kosack, K.E. and Jones, J.D.G., 1996. Resistance gene-dependent plant defense responses. *Plant Cell,* 8 (10), 1773-1791.

Harborne, J.B. (ed.) 1989. *Plant phenolics*. Academic Press, London. Methods in Plant Biochemistry no. 1.

Huala, E., Oeller, P.W., Liscum, E., et al. 1997. *Arabidopsis* NPH1: A protein kinase with a putative redox-sensing domain. *Science,* 278 (5346), 2120-2123.

Inbaraj, J.J. and Chignell, C.F., 2004. Cytotoxic action of juglone and plumbagin: a mechanistic study using HaCaT keratinocytes. *Chemical Research in Toxicology,* 17 (1), 55-62.

Inderjit, 1996. Plant phenolics in allelopathy. *Botanical Review,* 62 (2), 186-202.

Jamison, D.S., 2003. *Genetic analysis of haustorium formation in the parasitic plant Triphysaria.* PhD Thesis, University of California-Davis, Davis. Genetics.

Jamison, D.S. and Yoder, J.I., 2001. Heritable variation in quinone-induced haustorium development in the parasitic plant *Triphysaria. Plant Physiology,* 125 (4), 1870-1879.

Jones, E., Ekundayo, O. and Kingston, D.G.I., 1981. Plant anticancer agents. XI. 2,6-Dimethoxybenzoquinone as a cytotoxic constituent of *Tibouchina pulchra. Journal of Natural Products,* 44 (4), 493-494.

Kamei, H., Koide, T., Kojima, T., et al. 1998. Inhibition of cell growth in culture by quinones. *Cancer Biotherapy and Radiopharmaceuticals,* 13 (2), 185-188.

Kuijt, J., 1969. *The biology of parasitic flowering plants*. University of California Press, Berkeley.

Lee, K.C. and Campbell, R.W., 1969. Nature and occurrence of juglone in *Juglans nigra. HortScience,* 4, 297-8.

Li, R.B., Bianchet, M.A., Talalay, P., et al. 1995. The three-dimensional structure of NAD(P)H:quinone reductase, a flavoprotein involved in cancer chemoprotection and chemotherapy: mechanism of the two-electron reduction. *Proceedings of the National Academy of Sciences of the United States of America,* 92 (19), 8846-8850.

Losner-Goshen, D., Portnoy, V.H., Mayer, A.M., et al. 1998. Pectolytic activity by the haustorium of the parasitic plant *Orobanche* L. (*Orobanchaceae*) in host roots. *Annals of Botany,* 81 (2), 319-326.

Matúšová, R. and Bouwmeester, H.J., in press. The effect of host-root-derived chemical signals on the germination of parasitic plants. *In:* Dicke, M. and Takken, W. eds. *Chemical ecology: from gene to ecosystem.* Springer, Dordrecht. Wageningen UR Frontis Series no. 16. [http://library.wur.nl/frontis/chemical_ecology/04_matusova.pdf]

Matvienko, M., Torres, M.J. and Yoder, J.I., 2001a. Transcriptional responses in the hemiparasitic plant *Triphysaria versicolor* to host plant signals. *Plant Physiology,* 127 (1), 272-282.

Matvienko, M., Wojtowicz, A., Wrobel, R., et al. 2001b. Quinone oxidoreductase message levels are differentially regulated in parasitic and non-parasitic plants exposed to allelopathic quinones. *Plant Journal,* 25 (4), 375-387.
Muller, C.H., Muller, W.H. and Haines, B.L., 1964. Volatile growth inhibitors produced by aromatic shrubs. *Science,* 143, 471-3.
Nickrent, D., 2005. *Parasitic plant connection.* Available: [http://www.parasiticplants.siu.edu/] (24 Nov 2005).
Nishina, A., Hasegawa, K.I., Uchibori, T., et al. 1991. 2,6-dimethoxy-P-benzoquinone as an antibacterial substance in the bark of *Phyllostachys-heterocycla* var. *pubescens*, a species of thick-stemmed bamboo. *Journal of Agricultural and Food Chemistry,* 39 (2), 266-269.
Noda, Y., Kawazoe, Y. and Hakura, A., 1997. Cytotoxicity of naphthoquinones toward cultured resting murine leukemia L1210 cells in the presence of glutathione, diethyl maleate, or iodoacetamide. *Biological and Pharmaceutical Bulletin,* 20 (12), 1250-1256.
O'Brien, P.J., 1991. Molecular mechanisms of quinone cytotoxicity. *Chemico-Biological Interactions,* 80 (1), 1-41.
Parker, C. and Riches, C.R., 1993. *Parasitic weeds of the world: biology and control.* CAB International, Wallingford.
Press, M.C. and Graves, J.D. (eds.), 1995. *Parasitic plants.* Chapman & Hall, London.
Rao, P.V., Krishna, C.M. and Zigler, J.S., 1992. Identification and characterization of the enzymatic activity of zeta-crystallin from guinea pig lens: a novel NADPH quinone oxidoreductase. *Journal of Biological Chemistry,* 267 (1), 96-102.
Rhee, S.Y., Beavis, W., Berardini, T.Z., et al. 2003. The *Arabidopsis* Information Resource (TAIR): a model organism database providing a centralized, curated gateway to *Arabidopsis* biology, research materials and community. *Nucleic Acids Research,* 31 (1), 224-228.
Riopel, J.L. and Timko, M.P., 1995. Haustorial initiation and differentiation. *In:* Press, M.C. and Graves, J.D. eds. *Parasitic plants.* Chapman & Hall, London, 39-79.
Siqueira, J.O., Nair, M.G., Hammerschmidt, R., et al. 1991. Significance of phenolic compounds in plant-soil-microbial systems. *Critical Reviews in Plant Sciences,* 10 (1), 63-131.
Smith, C.E., Ruttledge, T., Zeng, Z.X., et al. 1996. A mechanism for inducing plant development: the genesis of a specific inhibitor. *Proceedings of the National Academy of Sciences of the United States of America,* 93 (14), 6986-6991.
TAIR, 2005. *Gene ontology at TAIR (The Arabidopsis Information Resource).* Available: [http://www.arabidopsis.org/tools/bulk/go/index.jsp] (24 Nov 2005).
Testa, B., 1995. *The metabolism of drugs and other xenobiotics: biochemistry of redox reactions.* Academic Press, New York.
Thorn, J.M., Barton, J.D., Dixon, N.E., et al. 1995. Crystal structure of *Escherichia coli* QOR quinone oxidoreductase complexed with NADPH. *Journal of Molecular Biology,* 249 (4), 785-799.
Thurman, L.D., 1966. *Genecological studies in Orthocarpus subgenus Triphysaria (Scrophulariaceae).* PhD Thesis, University of California, Berkeley.
Tomilov, A., Tomilova, N. and Yoder, J.I., 2004. In vitro haustorium development in roots and root cultures of the hemiparasitic plant *Triphysaria versicolor. Plant Cell, Tissue and Organ Culture,* 77 (3), 257-265.
Williamson, G.B., 1990. Allelopathy, Koch's postulates, and the neck riddle. *In:* Grace, J.B. and Tilman, D. eds. *Perspectives on plant competition.* Academic Press, San Diego, 143-162.
Wrobel, R.L., Matvienko, M. and Yoder, J.I., 2002. Heterologous expression and biochemical characterization of an NAD(P)H: quinone oxidoreductase from the hemiparasitic plant *Triphysaria versicolor. Plant Physiology and Biochemistry,* 40 (3), 265-272.
Yoder, J.I., 1997. A species-specific recognition system directs haustorium development in the parasitic plant *Triphysaria* (Scrophulariaceae). *Planta,* 202 (4), 407-413.
Yoder, J.I., 1998. Self and cross-compatibility in three species of the hemiparasite *Triphysaria* (Scrophulariaceae). *Environmental and Experimental Botany,* 39 (1), 77-83.
Zeng, Z.X., Cartwright, C.H. and Lynn, D.G., 1996. Chemistry of cyclopropyl-p-benzoquinone: a specific organogenesis inhibitor in plants. *Journal of the American Chemical Society,* 118 (5), 1233-1234.

CHAPTER 6

THE CHEMOSENSORY SYSTEM OF *CAENORHABDITIS ELEGANS* AND OTHER NEMATODES

DAMIEN M. O'HALLORAN, DAVID A. FITZPATRICK AND ANN M. BURNELL[#]

Institute of Bioengineering and Agroecology, Department of Biology, National University of Ireland Maynooth, Maynooth, Co. Kildare, Ireland.
[#] Corresponding author. E-mail: ann.burnell@nuim.ie

Abstract. Olfactory systems allow organisms to detect and discriminate between thousands of low molecular mass, mostly organic, compounds which we call odours. Organisms as diverse as humans and nematodes utilize the same basic mechanisms for this sensory perception. Represented in the olfactory repertoire of both vertebrates and invertebrates are aliphatic and aromatic compounds with diverse functional groups including aldehydes, esters, ketones, alcohols, ethers, carboxylic acids, amines, halides and sulphides. Soil-dwelling nematodes encounter many types of volatile and water-soluble molecules in their environment; successful foraging depends on the animal's ability to detect a gradient in one odorant while ignoring extraneous odours. Water-soluble chemicals tend to diffuse slowly in the soil and may provide short-range chemosensory cues whereas volatile compounds diffuse more rapidly and thus can be used for long-range chemotaxis to distant food sources. Animals modify their behaviour based on the interpretation of these environmental cues. The biochemical and physiological processes of chemosensory perception involve the recognition of small chemical molecules by specialized transduction pathways in the organism. These pathways are responsible for the transformation of information from extrinsic molecules into signals that the nervous system can interpret. The highly conserved G-protein signalling pathway is used to provide this chemosensory ability. The interaction of an odorant with an olfactory receptor results in the activation of heterotrimeric GTP-binding proteins (G proteins). G-protein signalling has been the subject of intense research over the last two decades. G proteins are present in all eukaryotic cells and signalling through G-protein-coupled receptors and heterotrimeric G proteins is one of the main means of transducing extracellular signals in the cell. *Caenorhabditis elegans* is an excellent model organism to study the molecular mechanisms behind signalling pathways in that it possesses unique traits amenable to both forward and reverse genetics. Exploiting these traits has shed much light on the mechanisms behind G-protein signalling. As molecular manipulations routinely used for *C. elegans* are becoming available for other nematodes, an increasing amount of chemosensory information is becoming available for a diverse range of nematodes from an even more diverse range of habitats.
Keywords: chemoreception; chemoreceptor genes; olfaction; nematode; G protein

M. Dicke and W. Takken (eds.), Chemical Ecology: From Gene to Ecosystem, 71-88.

NEMATODE CHEMOSENSORY BEHAVIOUR

Nematodes are thought to have diverged early in metazoan evolution (Poinar 1983). Diversity within the phylum Nematoda is enormous; there are nearly 20,000 species classified in the phylum. Nematodes occupy a wide range of habitats including terrestrial and marine environments. The vast majority are free-living microbivores, but many species have adopted a parasitic lifestyle. Most plants and animals have at least one parasitic nematode species uniquely adapted to exploit the concentration of food and resources that the host species represents. The relationships between nematodes and their hosts are also varied, so too are the reproductive strategies employed by nematodes. The development of adaptable sensory systems is central to survival. Through evolution, chemoreception has become the primary neurosensory tool used by nematodes to detect food sources, potential hosts, noxious compounds, reproductive partners and sometimes to enable them to choose between alternative developmental states (Krieger and Breer 1999; Prasad and Reed 1999).

The chemotactic responses of the free-living soil nematode *Caenorhabditis elegans* have been extensively investigated for over thirty years. *C. elegans* responds to a wide spectrum of water-soluble and volatile chemicals. Na^+, Li^+, Cl^- and OH^- ions are attractive to *C. elegans,* as are the water-soluble molecules cAMP, cGMP, lysine, histidine, cysteine and biotin (Ward 1973; Dusenbery 1974; Bargmann and Horvitz 1991). In the soil *C. elegans* feeds on a large variety of bacteria associated with decaying organic matter (Andrew and Nicholas 1976). The by-products of bacterial metabolism include various carboxylic acids, alcohols, aldehydes, esters, ketones and hydrocarbons (Zechman and Labows 1985; Schöller et al. 1997) and several of these compounds are highly attractive to *C. elegans* (Bargmann et al. 1993). In the aroma-rich soil environment, the infective stages of animal- and plant-parasitic nematodes need to be able to detect diagnostic host-specific odours to enable them to locate and infect appropriate hosts. Carbon dioxide is a well characterized attractant which is produced as an end product of metabolism by plants, micro-organisms and animals. The plant parasitic nematode, *Meloidogyne incognita*, has been shown to respond to a gradient of carbon dioxide (Pline and Dusenbery 1987). Using cylinders of moist sand Robinson (1995) showed that *M. incognita*, *Rotylenchulus reniformis* and *Steinernema glaseri* were all attracted to a linear gradient of carbon dioxide. Numerous free-living marine nematodes aggregate in and around decaying animal bodies and plant material. Riemann and Schrage (1988) demonstrated that the free-living marine nematode Adoncholaimus *thalassophygas* was attracted to carbon dioxide, which may help it to locate sites of anaerobic decomposition as a source of food.

Unlike free-living nematodes such as *C. elegans*, which feed on a wide range of bacterial species (Andrew and Nicholas 1976; Balan 1985) as well as filaments of fungal mycelium, fungal spores and yeast (Balanova and Balan 1991), parasitic nematodes must fine-tune their chemosensory repertoire to respond more precisely to host-specific cues. Plant-parasitic nematodes respond to plant allelochemicals to ensure close synchrony between host and parasitic life cycles. The majority of plant-parasitic nematodes infect plant roots and some have evolved sophisticated interactive relationships with host cells to sustain a sedentary parasitic habit. The root-knot nematodes, *Meloidogyne* spp., have a potential host range encompassing more than 3000 plant species. Potato root diffusates stimulate movement of hatched juveniles of *Globodera rostochiensis* (Clarke and Hennessy 1984) and may aid in host location. However, exposure of males of *G. rostochiensis* to the diffusate elicits no response (Riga et al. 1996). Males exit from the roots into the soil but probably remain in close proximity to the roots, apparently needing only sex pheromones to attract them to females. Masamune et al. (1982) isolated a natural hatching stimulus for the soybean cyst nematode. This stimulus, called glycinoeclepin A, was shown to stimulate the hatching of larvae from eggs *in vitro* from the roots of kidney beans. Although root diffusates are generally considered attractive to nematodes, several chemicals produced within the roots of some plants have been characterized that repel plant-parasitic nematodes. One such plant is the marigold (*Tagetes* spp.), which produces the compound α-terthienyl (Bakker et al. 1979; Gommers and Bakker 1988). This compound when photoactivated produces reactive oxygen species, which are highly toxic to nematodes. The compound, α-terthienyl, has been used to suppress populations of certain economically important plant-parasitic nematodes.

Entomopathogenic nematodes (EPNs) are a ubiquitous group of obligate and lethal parasites of insects. They are characterized by their ability to carry and transmit a specific insect-pathogenic symbiont bacterium. Two EPN families are currently recognized: the Steinernematidae and the Heterorhabditidae. Analysis of small-subunit ribosomal DNA reveals that these families are not closely related phylogenetically (Blaxter et al. 1998), but appear to have evolved similar morphological and ecological traits through convergent evolution (Poinar 1983). As parasitic nematodes have a more focused life cycle than free-living nematodes it is not surprising that the insect-parasitic nematode, *Heterorhabditis bacteriophora* has a similar but more restricted chemosensory repertoire than that of the free-living nematode, *C. elegans* (O'Halloran and Burnell 2003). The most notable difference in the chemotactic responses of these two nematode species is that *H. bacteriophora* infective juveniles are unresponsive to a large number of compounds which *C. elegans* finds highly attractive. The latter compounds are typical by-products of bacterial metabolism and include aldehydes, esters, ketones and short-chain alcohols (Bargmann et al. 1993), which would not provide helpful cues to assist a parasitic nematode find its host. Rasmann et al. (2005) reported the first identification of an insect-induced below-ground plant signal, (*E*)-β-caryophyllene, which strongly attracts the EPN, *Heterorhabditis megidis*. This plant signal is a sesquiterpene released by maize roots in response to feeding by the larvae of the beetle *Diabrotica virgifera virgifera*. (*E*)-β-caryophyllene is only detected from maize leaves and roots after herbivory and so is probably not the only long-range attractant for *H. megidis*, as Rasmann et al. (2005) also demonstrated that nematodes were moderately attracted to healthy and mechanically damaged plants. Therefore, what we see is that many species of nematode are adapted to a very specific repertoire of odours, which are used to exploit the concentration of food and resources that the host or food source represents.

THE CHEMOSENSORY SYSTEM OF *C. ELEGANS* AND OTHER NEMATODES

The nematode nervous system is designed to integrate many distinct environmental stimuli so that the organism can respond appropriately. A great deal is known about the properties of the neuronal circuits and the specialized neurons that encode sensory information in *C. elegans* (White et al. 1978; 1986; 1991). The *C. elegans* brain consists of a circumpharyngeal nerve ring containing 302 neurons comprising 118 morphologically distinct cell types, all of which interconnect in a reproducible manner to form a variety of neural circuits and pathways. Gap junctions occur between neurons and between muscle cells. *C. elegans* neurons have a simple (mostly monopolar or dipolar), relatively unbranched morphology and nerve processes are generally organized into ordered bundles, which, in the majority of classes, run longitudinally (e.g., ventral and dorsal nerve cords) or circumferentially (commissures).

In *C. elegans* processes from the circumpharyngeal nerve ring run anteriorly as six cephalic (head region) nerve bundles forming the inner and outer labial neurons. The dendrites in four of these nerve bundles have their cell bodies just anterior to the nerve ring, in a region loosely referred to as the 'anterior ganglion' (Chalfie and White 1988). Axons from these cell bodies synapse with the nerve ring. Two other cephalic nerve bundles contain processes from the lateral ganglia, from which amphid neuronal axons run into the nerve ring. Two bilaterally symmetric amphids in the *C. elegans*' head each contain the dendritic endings of 12 types of sensory neurons.

The nematode chemosensory organs are the amphids (Figure 1), located near the head, and the phasmids (Figure 2), located at the nematode's posterior. Nematodes are subdivided into two classes by presence or absence of phasmids, the Class Secernentea which has phasmids and the Class Adenophorea which does not possess phasmids. Phasmids are similar in general structure to the amphids, both consisting of a group of neurons opening to the exterior. In *C. elegans*, chemosensory cells within the phasmid negatively modulate reversals to repellents (Hilliard et al. 2002). The amphid neurons responsible for chemosensory and thermosensory behaviours have been identified in *C. elegans* (Secernentea) through behavioural analysis of animals in which defined neurons were ablated using a laser microbeam (Bargmann and Mori 1997). Eight types of neurons (ADF, ADL, ASE, ASG, ASH, ASI, ASJ, ASK) have one or two long slender cilia that are directly exposed to the environment through the amphid pore (Figure 1). These neurons detect mostly water-soluble chemicals (Table 2). Three types of neurons (AWA, AWB, AWC) have flattened, branched cilia that terminate near the amphid pore, but enclosed by a support cell called the amphid sheath cell. These neurons detect volatile odorants (Table 2). One type of neuron that detects thermal cues (AFD) has a complex, brush-like dendritic membrane structure at the sensory ending which is embedded in the amphid sheath cell (Figure 1).

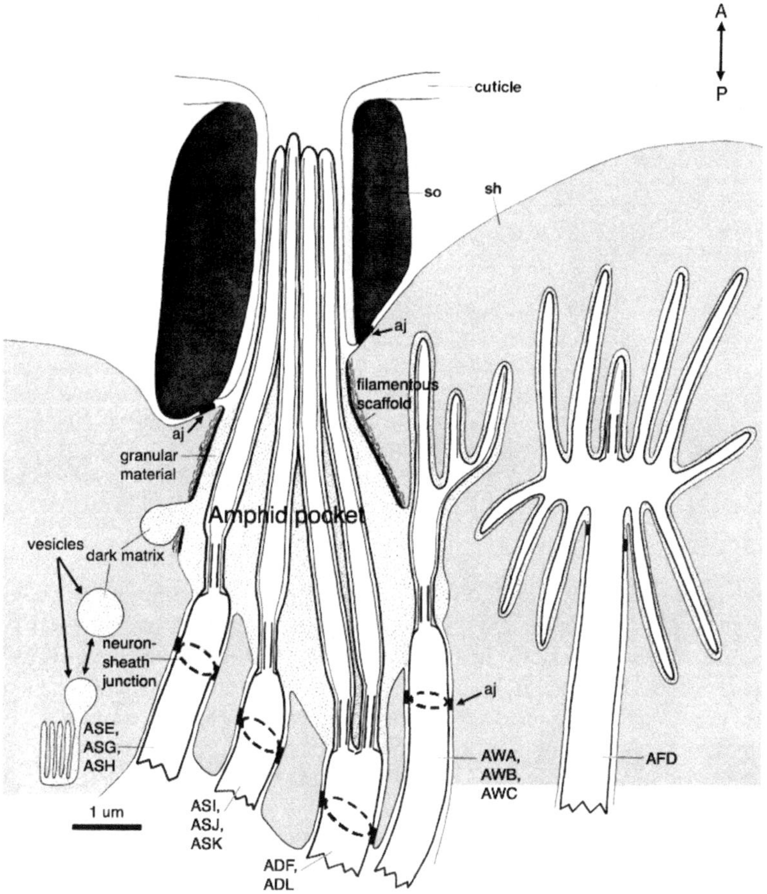

Figure 1. *Schematic longitudinal section through an amphid of C. elegans. The amphid channel is formed from a socket cell (so) and a sheath cell (sh). The socket cell is joined by belt junctions to surrounding hypodermal cells. The socket channel is lined with cuticle that is continuous with the external cuticle. The anterior sheath channel has a dark, non-citicular lining surrounded by a filamentous scaffold. The sheath and socket cells are joined together by belt junctions encircling the channel. The space between the cilia in the posterior sheath channel is filled with a dark matrix that appears to be packaged into vesicles further posterior, transported forward, and deposited around the cilia. The dendrites of three channel neurons and one wing neuron (AWA) are shown. The distal segment of the AWA cilium leaves the fascicle of channel cilia to re-invaginate the sheath cell. The AFD dendrite remains separate from the fascicle of wing and channel cilia. All of the dendrites form belt-shaped junctions with the sheath cell near their point of invagination. Main scale bar is 1.0 micrometer and A P arrows refer to anterior and posterior direction. (Reproduced with permission from www.wormatlas.org)*

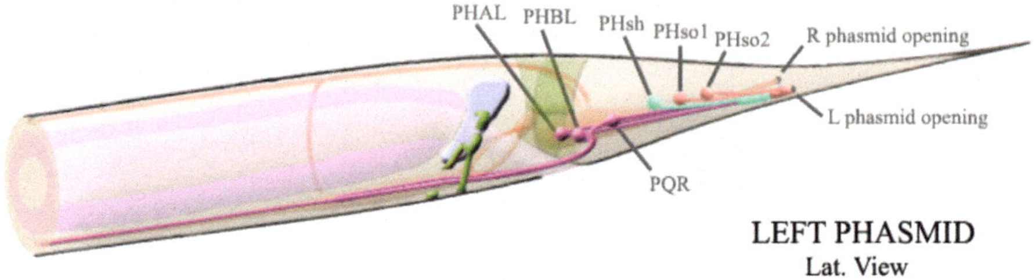

Figure 2. *Illustration of the lateral view of the left phasmid of C. elegans. The phasmids are similar in their structure to amphid sensilla, but smaller. They are located at the lateral sides of the tail and enclose the ciliated dendrites of PHA, PHB and, on the left side, PQR neurons as well as one sheath (sh) and two socket cells (so1 and so2). The cilia of the PHA and PHB neurons extend into the external medium through the channel created by the socket cells. The ending of posterior process of PQR is wrapped by PHso2L. Phasmid sheath cells extend short processes posteriorly into tail tip which swell to form a protective pocker near the phasmid openings for PHA and PHB cilia. (Reproduced with permission from www. wormatlas.org)*

There is considerable variation in size and form of the amphids between the Secernentea and the Adenophorea. Typically, paired amphids are situated laterally, but in some Adenophorea and in many Secernentea the amphids are more dorsal. The Adenophorea display much variation in their amphid organs and adenophorean amphids are usually larger and often present in greater numbers than are secernentean amphids. The microbivore *Leptonemella* spp. is a member of the Adenophorea, with large amphids (18-30 μm long) that display sexual dimorphism in their morphology, being spiral in females and loop-shaped in males (Hoschitz et al. 1999). In several animal-parasitic nematodes belonging to the Secernentea the positions of the amphidial neuronal cell bodies in the lateral ganglia are analogous to that observed in *C. elegans* (Ashton et al. 1999). Because positional homologies are conserved between these nematodes species it is likely that many functional homologies are also conserved. Ashton et al. (1999) investigated two neuron classes (ASF and ASI) in the parasitic nematode *Strongyloides stercoralis*. They found that these neurons control the decision whether to become an infective larva directly (homogonic development) or to become a free-living adult worm. This developmental switch parallels the decision in C. elegans whether to become a dauer larva (when conditions are adverse) or to continue normal development to adulthood. In the same study Ashton et al. (1999) noted that the ASE class of amphidial neurons in *S. stercoralis* had a chemosensory function, as in *C. elegans*, but unlike *C. elegans* this same neuron also has a thermosensory function.

Table 1. *Responses of H. bacteriophora to volatile and water-soluble compounds (O'Halloran and Burnell 2003).*

Attractants	
Alcohols	1-pentanol*, 1-hexanol*, 1-heptanol, 2-heptanol, 1-octanol, 2-octanol, 1-nonanol, 2-nonanol, 3-nonanol
Thiazole/Pyrazine	4,5-dimethylthiazole, 2-isobutylthiazole, 2-methylpyrazine, benzothiazole, 2-acetylthiazole
Organic acids	caproic acid, caprylic acid, methylvaleric acid
Others	carbon dioxide, dry-ice
Weak attractants	
Alcohols	2-mercaptoethanol, 1-butanol, 1-propanol, 1-ethanol, 3-heptanol
Others	carbonated water, uric acid¶, host assay, hexanal
Neutral compounds	
Alcohols	isobutanol, isoamyl alcohol
Ketones	acetone, 2-butanone, 2-pentanone, 2-hexanone, 2-heptanone, diacetyl
Aldehydes	benzaldehyde, valeraldehyde
Pyrazines	acetylpyrazine
Amines	butylamine
Esters	ammonium acetate, isopropyl acetate, isoamyl acetate, ethyl acetate
Others	copper sulphate¶, L-cysteine¶, dimethyl sulphoxide, paraffin, formamide, zinc sulphate¶, diethyl ether
Repellents	
Alcohols	methanol, 1-hexanol*, 1-pentanol*
Pyrazines	2,6-dimethylpyrazine, pyrazinamide
Others	L-lysine¶, d-biotin¶

*Some molecules listed with an asterisk are attractive at high concentrations and repellent at low concentrations.

¶These compounds were applied to the agar 120 minutes before the infective juveniles were added.

Table 2. *Neuronal functions in C. elegans as defined by laser ablation (Bargmann and Mori 1997).*

	Neuron	Function
Sensory neurons	AWA	volatile chemotaxis; diacetyl, pyrazine, thiazole
	AWB	volatile avoidance
	AWC	volatile chemotaxis; benzaldehyde, butanone, isoamyl alcohol, thiazole
	AFD	thermotaxis
	ASE	Na^+, Cl^-, cAMP, biotin, lysine chemotaxis, egglaying
	ADF	dauer pheromone; Na^+, Cl^-, cAMP, biotin chemotaxis
	ASG	dauer pheromone; Na^+, Cl^-, cAMP, biotin, lysine chemotaxis
	ASH	osmotic avoidance, nose-touch avoidance, volatile avoidance
	ASI	dauer pheromone; Na^+, Cl^-, cAMP, biotin, lysine chemotaxis
	ASJ	dauer pheromone (recovery)
	ASK	lysine chemotaxis, egg-laying
	ADL	octanol avoidance, water-soluble avoidance

MOLECULAR MECHANISMS OF CHEMOTAXIS

Chemoreceptor genes in Caenorhabditis

A variety of behavioural screens have been developed in *C. elegans* to identify mutant nematodes with defects in their chemosensory behaviours. These include direct screens for chemotaxis-defective mutants (che and tax – Ward 1973; Dusenbery 1974) as well as nematodes with defective responses to volatile odorants, but not to water-soluble attractants (*odr* mutants – Bargmann et al. 1993). Some chemosensory neurons are involved in dauer formation and so some chemosensory mutants were first isolated based upon defects in their ability to form dauer larvae (daf mutants, e.g., daf-11 and daf-21 – Riddle et al. 1981; Thomas et al. 1993).

The first chemoreceptor genes in C. elegans were isolated using a bioinformatics approach (Troemel et al. 1995). A cluster of 9 related genes were found adjacent to a transmembrane guanylyl cyclase and these genes encoded proteins with multiple predicted transmembrane domains. These sequences were then used to search the C. elegans genome for related genes, and a total of 41 putative receptor genes representing 6 families sra, srb, srd, sre, srg and sro were identified (sr = serpentine receptor, a term sometimes used for 7-TM receptors). Of 14 genes for which expression data were obtained, eleven were expressed only in small subsets of chemosensory neurons. The low levels of similarity within these 7-TM sub-families explain the small number of genes identified via this approach. For example, the three largest families of genes identified by Troemel et al. (1995) were the sra, srb and srg genes. The sra family shared only about 35% amino-acid identity overall, the eleven srb genes were distantly related from the sra genes and shared only 10-15% amino-acid similarity. The thirteen srg genes identified were essentially unrelated to the sra or srb genes by sequence, but were between 10 and 30% similar to one another. When the odr-10 gene was cloned (Sengupta et al. 1996) it was found to be a divergent 7-TM receptor with a weak homology to the srd genes identified by Troemel et al. (1995), and it also had a weak similarity to vertebrate olfactory receptors (~10% amino-acid identity). Odr-10 mutants were isolated from C. elegans in behavioural screens for animals that failed to respond to the odorant diacetyl (Sengupta et al. 1996). The odr-10 gene is expressed only in the cilia of the AWA olfactory neurons in each amphid. Mutations in the odr-10 gene lead to a selective loss in the nematodes' ability to sense diacetyl, however the nematodes exhibit normal chemotaxis to other odorants recognized by the AWA olfactory neurons, and thus are not completely defective in AWA function. odr-10 cDNA also specifically restores diacetyl sensitivity to mutants that have lost their ability to respond to several odorants (such as odr-7, which have defective expression of a transcription factor controlling odr-10 expression, Sengupta et al. 1996). The function of ODR-10 as a chemoreceptor was further confirmed when odr-10 was transformed into mammalian cells where it functioned as a diacetyl-activated chemoreceptor (Zhang et al. 1997).

Unlike vertebrate genes encoding olfactory receptors, the *odr-10* gene contains introns (Robertson 1998). The sequence similarity between ODR-10 and the vertebrate olfactory receptors is limited to a few residues in the predicted proteins; however, these two receptor families do share more similarity with each other than with other G-protein-linked receptors. Nevertheless it is difficult to discern whether vertebrate and invertebrate olfactory receptors are derived from a common ancestor (Robertson 2000; 2001). Analysis of the *C. elegans* genome by Robertson (Robertson 1998; 2001) suggests that it may encode ~550 functional chemoreceptor genes and ~250 pseudogenes, which together represent ~6% of the genome. There is an ongoing and rapid process of gene duplication, deletion, diversification and movement in nematode chemoreceptor genes. For example, comparison with the *C. briggsae* genome indicates that ~28% of the *C. elegans srh* 7-TM family have been newly formed since the split with *C. briggsae* (Robertson 2001). Another point of interest is the significant reduction in chemoreceptor genes in the *C. briggsae* genome. The *srz* chemoreceptor family has 60 representatives in the *C. elegans* genome compared with only 28 members within the *C. briggsae* genome (Thomas et al. 2005). The *srz* family also displays frequent gene duplication and deletion events as well as possessing sites undergoing positive selection (Thomas et al. 2005). The chemoreceptor subfamily five has 311 members in *C. elegans* and only 151 representatives in *C. briggsae.* Also, the *sra* family of chemoreceptors has 36 and 18 members in *C. elegans* and *C. briggsae*, respectively (Stein et al. 2003). Overall, *C. briggsae* has over 40% fewer chemoreceptor genes than *C. elegans*, highlighting the rapid rate of evolution of the chemoreceptor gene family in these nematodes.

Heterotrimeric G protein subunits

The heterotrimeric guanine nucleotide-binding proteins (G proteins) act as switches that regulate information-processing circuits connecting cell-surface receptors to a variety of effectors such as nucleotide cyclases and ion channels. The G proteins are present in all eukaryotic cells and control metabolic, humoral, neural and developmental functions. In animals as different as humans and worms, G proteins mediate olfactory discrimination (Prasad and Reed 1999). G proteins are comprised of three peptides: an α subunit that binds and hydrolyses guanosine triphosphate (GTP), a β subunit and a γ subunit. The β and γ subunits form a dimer that only dissociates when it is denatured, thus representing a functional monomer.

When GDP is bound, the α subunit associates with the βγ subunit to form an inactive heterotrimer that binds to the receptor (Figure 3). Both α and βγ subunits can bind to the receptor. Monomeric, GDP-liganded α subunits can interact with receptors, but the association is greatly enhanced in the αβγ heterotrimer. When a chemical or physical signal stimulates the receptor, the receptor becomes activated and changes its conformation. The GDP-liganded α subunit responds with a conformational change that decreases GDP affinity, so that GDP comes off the active site of the α subunit (Figure 3). Because the concentration of GTP in cells is much higher than that of GDP, the outgoing GDP is replaced with GTP. Once GTP is bound, the α subunit assumes its activated conformation and dissociates both from the receptor and from βγ. The activated state lasts until the GTP is hydrolysed to GDP by the intrinsic GTPase activity of the α subunit. All isoforms of α subunits are GTPases, although the intrinsic state of GTP hydrolysis varies greatly from one type of Gα subunit to another (Carty et al. 1990; Linder et al. 1990). Once GTP is cleaved to GDP, the α and βγ subunits reassociate, the heterotrimer becomes inactive and returns to the receptor. The free α and βγ subunits each activate target effectors. Figure 3 illustrates the cycle of G-protein activation and deactivation that transmits a signal from receptor to effector. Six Gβ and twelve Gγ gene products have been identified in mammals (Hamm 1998). In the *C. elegans* genome two Gβ genes and two Gγ genes have been identified (Van der Voorn et al. 1990; Jansen et al. 1999).

GTP-binding α subunits have been divided on the basis of amino-acid similarity into four classes in mammals; $G\alpha_s$, $G\alpha_i$, $G\alpha_q$ and $G\alpha_{12}$. Each grouping has been shown to function differently. Subunits of the $G\alpha_{12}$ class were originally isolated from a mouse-brain cDNA library (Strathmann and Simon 1991) and since have been shown to be expressed ubiquitously in diverse cell lines and tissues from different species (Dhanasekaran and Dermott 1996). Similarly, members of the $G\alpha_s$ and $G\alpha_{i/o}$ classes have been shown to be expressed in a wide range of tissue types. In *C. elegans*, *gsa-1* and *goa-1* (homologues of the $G\alpha_s$ and $G\alpha_{i/o}$ classes, respectively) were expressed in all cells examined (Jansen et al. 1999). The members of the $G\alpha_q$ class are often co-expressed in a variety of cell types (Milligan et al. 1993). The $G\alpha_s$ class stimulates cAMP production (Graziano et al. 1987), in contrast to $G\alpha_i$ proteins, which inhibit cAMP production and are sensitive to the Pertussis toxin (PTX) (Simon et al. 1991). $G\alpha_q$ proteins have been shown to be refractory to PTX modification (Simon et al. 1991) and the $G\alpha_{12}$ class represents yet another class of PTX-insensitive Gα proteins (Parks and Wieschaus 1991).

In *C. elegans* a representative of each of the four main mammalian Gα classes is present, as well as 17 additional Gα subunits, giving a total of 21 GTP-binding Gα subunit genes (Jansen et al. 1999). Fourteen of the additional Gα subunits are expressed almost exclusively in a small subset of the chemosensory neurons in

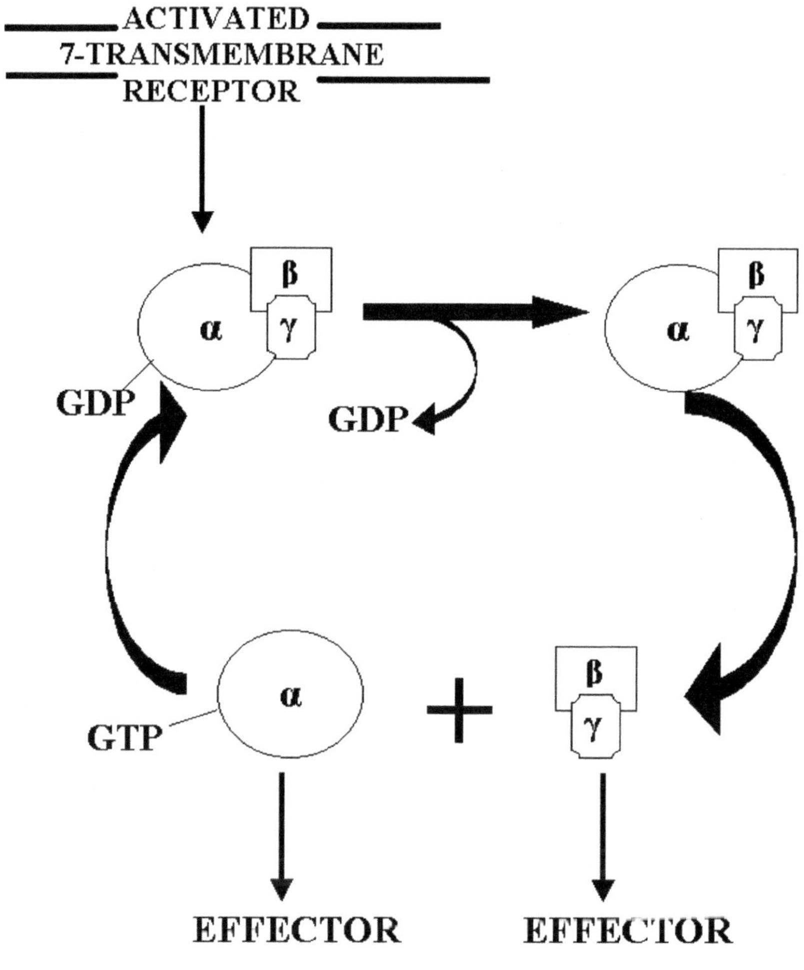

Figure 3. *The regulatory cycle of heterotrimeric G proteins subunits. When a chemical or physical signal stimulates the receptor, the receptor becomes activated and changes its conformation. The GDP-liganded a subunit responds with a conformational change that decreases GDP affinity, so that GDP comes off the active site of the a subunit and is replaced with GTP. Once GTP is bound, the a subunit assumes its activated conformation and dissociates both from the receptor and from βγ. The activated state lasts until the GTP is hydrolysed to GDP by the intrinsic GTPase activity of the α subunit. Once GTP is cleaved to GDP, the α and βγ subunits reassociate, the heterotrimer becomes inactive and returns to the receptor. The free α and βγ subunits each activate target effectors. Black lines indicate the neuronal membrane*

C. elegans (Jansen et al. 1999). Although none of the Gα genes expressed in *C. elegans* amphids are essential for viability, their expression pattern clearly indicates a role for them in chemoreception. Similarly to chemosensory receptors, multiple Gα subunit genes are used in each cell (Jansen et al. 1999). We have constructed a data set containing homologues of putative Gα genes from a variety of metazoa, protistans and fungi. The final alignment contained 146 taxa and 751 aligned amino-acid positions. Our analysis reveals that nematodes have evolved multiple novel Gα subunit genes through a series of duplication events early in nematode evolution (O'Halloran et al. unpublished data). A single *C. elegans* olfactory neuron expresses multiple chemoreceptors and multiple heterotrimeric G proteins (Troemel et al. 1995; Jansen et al. 1999). The novel nematode-specific Gα genes increase the functional complexity of individual chemosensory neurons and facilitate the integration of signals from different odorant molecules within a single neuron.

Downstream signalling from chemoreceptors and G proteins

G-protein-mediated signalling is intrinsically kinetic. Signal amplitude is determined by the balance of the rates of GDP/GTP exchange (activation) and of the rates of GTP hydrolysis (deactivation). Downstream of G proteins, several novel proteins implicated in the deactivation and activation processes of Gα proteins have come into light in recent years (Ross and Wilkie 2000). Proteins involved in the deactivation process have been termed GTPase-activating proteins (GAPs) and include the G_q-stimulated phospholipase C-β (PLC-β) and the mammalian G_{13}-stimulated p115RhoGEF, a guanine nucleotide exchange factor for Rho GTPase (Chen et al. 2001). The most recently identified regulators of G-protein-signalling (RGS) proteins are found throughout most eukaryotes and are also G-protein GAPs (Watson et al. 1996).

RGS proteins accelerate the GTPase activity of G-protein α-subunits, thus driving them to their native inactive state. Mammals have ~20 proteins containing the ~120 amino-acid RGS domain that defines RGS proteins. The RGS domain folds into a nine-helix structure that binds to the Gα subunit, thereby stimulating its GTPase activity (Tesmer et al. 1997). Although many RGS proteins consist of little more than an RGS domain, a subset of them also contain a large amino-terminal conserved region of unknown function, as well as a G gamma-like (GGL) domain that is able to bind a specific G-protein β subunit (Snow et al. 1998; Chase et al. 2001). Thirteen RGS genes have been identified in *C. elegans.* Two of these have been analysed and shown to act on the homologues of the G proteins G_o and G_q (known as GOA-1 and EGL-30, respectively). The RGS protein EGL-10 inhibits signalling by G_o, which in turn inhibits egg-laying and locomotion behaviours (Mendel et al. 1995), whereas the RGS protein EAT-16 inhibits signalling by G_q, which has effects that are the opposite of those caused by G_o (Brundage et al. 1996; Miller et al. 1999). EGL-10 and EAT-16 are the only two RGS proteins in nematodes with GGL domains and have been shown to bind Gβ *in vivo*, although it is still unclear how this might influence RGS activity.

Other downstream components of the chemosensory network of *C. elegans* have been described, such as the *odr-1* and *daf-11* genes, which code for guanylyl cyclase, an effector enzyme responsible for producing the secondary messenger (cGMP) via heterotrimeric G-proteins. Guanylyl cyclase expression is essential for all AWC-sensed odorants (L'Etoile and Bargmann 2000). The heteromeric TAX-2/TAX-4 cyclic-nucleotide gated cation channel is sensitive to cGMP and insensitive to cAMP, suggesting that *C. elegans* uses cGMP as a second messenger in olfaction, unlike mammals, which have been shown to utilize cAMP as a secondary messenger in olfactory neurons (Prasad and Reed 1999). Another novel protein required for olfaction, mechanosensation and olfactory adaptation in *C. elegans* is OSM-9, a multiple transmembrane domain protein required for the activity of ODR-10 (Colbert et al. 1997). Bioinformatic analyses of *osm-9* revealed a previously unsuspected diversity of mammalian and invertebrate genes in this family. Cyclic-nucleotide gated-channel mutants such as *tax-2* or *tax-4* respond normally to some olfactory stimuli suggesting an alternative pathway of chemosensation which may involve *osm-9* (Colbert et al. 1997). Other olfactory effectors downstream of the receptor include various kinases. EGL-4 is a cGMP-dependent kinase which regulates multiple developmental and behavioural processes (Fujiwara et al. 2002; L'Etoile et al. 2002). The classical Ras-MAPK (mitogen-activated protein kinase) signal transduction pathway was also shown to be activated in *C. elegans* upon application of the attractant isoamyl alcohol (Hirotsu et al. 2000). Thus it is clear that G-protein-coupled odour transduction pathways are complex in mammalian systems, but are more complex still in nematodes in which multiple signal transduction mechanisms in the same cell are used to distinguish between odorants.

FUTURE PROSPECTS

The molecular tools that have been used to investigate the chemosensory system of *C. elegans* are now being developed and applied to studies on other nematodes. This technology transfer of research methodology from *C. elegans* is a slow process because of the diversity of nematodes studied by researchers and the lack of resources devoted to individual systems. Many groups have exploited the molecular knowledge of *C. elegans* to study other nematode systems. Kwa et al. (1995) were one of the earliest groups to demonstrate the use of *C. elegans* to study parasitic-nematode genes. They designed a mutant rescue assay to show that the β-tubulin genes from *Haemonchus contortus* could modulate drug resistance in *C. elegans*. Another more recent study demonstrated the ectopic expression of an *H. contortus* GATA transcription factor (*elt-2*) in *C. elegans* (Couthier et al. 2004). This factor is a central regulator of endoderm development. This study showed that the development of the *H. contortus* lineage is strikingly similar to that of *C. elegans*. Transformation of *C. elegans* with promoter/reporter gene constructs for the pepsinogen gene, *pep-1*, from *H. contortus* and the cysteine protease gene (*ac-2*) in *Ostertagia circumcinta* has also been demonstrated by Britton et al. (1999), revealing good spatial agreement with the localization of the native proteins encoded by these genes in the parasites. Hashmi et al. (1998) had some success at transforming *H. bacteriophora* by microinjection of reporter constructs. This transformation resulted in approximately 7% of the F1 generation exhibiting *lacZ* expression. Urwin et al. (2002) demonstrated that ingestion of dsRNA by preparasitic juvenile cyst nematodes leads to RNA interference of cysteine proteinases, major sperm proteins and a novel *Heterodera glycines* gene. Taken together, these studies suggest a high degree of conservation of the post-transcriptional and post-translational gene regulatory mechanisms between parasitic nematodes and *C. elegans*. As research methods from *C. elegans* and indeed other model organisms too are utilized by nematode researchers, a substantial amount of genetic, phylogenetic and pharmacogenomic knowledge pertaining to olfaction is gradually coming to light.

It seems that genes implicated in the nematode nervous system often have peculiarities associated with them. Along with the expansion and diversification of some neuronal gene families there has been a selective reduction and/or loss of certain others. For example, the largest and most diverse nicotinic acetylcholine receptor (nACHR) gene family is that of *C. elegans* (Mongan et al. 1998). nACHRs mediate the fast actions of the neurotransmitter acetylcholine at nerve muscle junctions and in the nervous system. The molecular diversity within this family includes very distinct groups, which are thought to have diverged early in nematode evolution (Treinin and Chalfie 1995). Lineage-specific expansion of neural Gα genes also appears to have occurred in nematodes. The NGF family of neurotrophins are protein growth factors with crucial roles in the determination of neuronal survival and regulation of neuronal numbers throughout vertebrate development. Completion of the *C. elegans* genome sequence has confirmed that the distinct 'hard wired' nematode nervous system does not require neurotrophins or their receptors (Ruvkun and Hobert 1998). *C. elegans* lacks voltage-dependent sodium-channel genes, which are present in the more primitive jellyfish (Bargmann 1998), suggesting the ability to generate a sodium-based action potential was lost during nematode evolution. Conversely a novel *osm*-related transient receptor potential (TRP) ion channel, OSM-9, was identified in *C. elegans* (Colbert et al. 1997), revealing the existence of an alternative chemosensory pathway within the nematode. Whole genome analysis of *C. elegans* and *C. briggsae*, in conjunction with previous studies on the *C. elegans* nervous system, indicate that the *Caenorhabditis* nervous system has some very specialized features and there are several examples of neuronal gene families that appear to have undergone nematode-specific gene expansion. Functional analysis of the *C. elegans* and *C. briggsae* gene sequences may possibly identify other novel gene families involved in nematode chemoreception. As additional nematode-sequencing projects are completed they will provide further whole-genome windows into the level of complexity associated with chemosensory signalling, as well as providing a platform of comparative genomics between a variety of divergent nematodes.

ACKNOWLEDGEMENTS

We are grateful to Cambridge University Press and Cold Spring Harbor Publishing for permission to publish Tables 1 and 2, respectively. Our research has been funded by the Irish Higher Education Authority PRTLI programme.

REFERENCES

Andrew, P.A. and Nicholas, W.L., 1976. Effect of bacteria on dispersal of *Caenorhabditis elegans* (Rhabditidae). *Nematologica,* 22 (4), 451-461.

Ashton, F.T., Li, J. and Schad, G.A., 1999. Chemo- and thermosensory neurons: structure and function in animal parasitic nematodes. *Veterinary Parasitology,* 84 (3/4), 297-316.

Bakker, J., Gommers, F.J., Nieuwenhuis, I., et al. 1979. Photoactivation of the nematocidal compound alpha-terthienyl from roots of marigolds (*Tagetes* spp): a possible singlet oxygen role. *Journal of Biological Chemistry,* 254 (6), 1841-1844.

Balan, J., 1985. Measuring minimal concentrations of attractants detected by the nematode *Panagrellus redivivus*. *Journal of Chemical Ecology,* 11 (1), 105-112.

Balanova, J. and Balan, J., 1991. Chemotaxis-controlled search for food by the nematode *Panagrellus redivivus*. *Biologia,* 46 (3), 257-263.

Bargmann, C.I., 1998. Neurobiology of the *Caenorhabditis elegans* genome. *Science,* 282 (5396), 2028-2033.

Bargmann, C.I., Hartweig, E. and Horvitz, H.R., 1993. Odorant-selective genes and neurons mediate olfaction in *C. elegans*. *Cell,* 74 (3), 515-527.

Bargmann, C.I. and Horvitz, H.R., 1991. Chemosensory neurons with overlapping functions direct chemotaxis to multiple chemicals in *C. elegans*. *Neuron,* 7 (5), 729-742.

Bargmann, C.I. and Mori, I., 1997. Chemotaxis and thermotaxis. *In:* Riddle, D.L., Blumenthal, T., Meyer, B.J., et al. eds. *C. elegans II*. Cold Spring Harbor Press, New York, 717-738.

Blaxter, M.L., De Ley, P., Garey, J.R., et al. 1998. A molecular evolutionary framework for the phylum Nematoda. *Nature,* 392 (6671), 71-75.

Britton, C., Redmond, D.L., Knox, D.P., et al. 1999. Identification of promoter elements of parasite nematode genes in transgenic *Caenorhabditis elegans*. *Molecular and Biochemical Parasitology,* 103 (2), 171-181.

Brundage, L., Avery, L., Katz, A., et al. 1996. Mutations in a *C. elegans* Gq alpha gene disrupt movement, egg laying, and viability. *Neuron,* 16 (5), 999-1009.

Carty, D.J., Padrell, E., Codina, J., et al. 1990. Distinct guanine nucleotide binding and release properties of the three g(i) proteins. *Journal of Biological Chemistry,* 265 (11), 6268-6273.

Chalfie, M. and White, J., 1988. The nervous system. *In:* Wood, W.B. ed. *The nematode Caenorhabditis elegans*. Cold Spring Harbour Press, New York, 337-391.

Chase, D.L., Patikoglou, G.A. and Koelle, M.R., 2001. Two RGS proteins that inhibit G[alpha](o) and G[alpha](q) signaling in *C. elegans* neurons require a G[beta](5)-like subunit for function. *Current Biology,* 11 (4), 222-231.

Chen, Z., Wells, C.D., Sternweis, P.C., et al. 2001. Structure of the rgRGS domain of p115RhoGEF: p115RhoGEF, a guanine nucleotide exchange factor for Rho GTPase, is also a GTPase. *Nature Structural Biology,* 8 (9), 805-809.

Clarke, A.J. and Hennessy, J., 1984. Movement of *Globodera rostochiensis* juveniles stimulated by potato-root exudate. *Nematologica,* 30 (2), 206-212.

Colbert, H.A., Smith, T.L. and Bargmann, C.I., 1997. OSM-9, a novel protein with structural similarity to channels, is required for olfaction, mechanosensation, and olfactory adaptation in *Caenorhabditis elegans*. *Journal of Neuroscience,* 17 (21), 8259-8269.

Couthier, A., Smith, J., McGarr, P., et al. 2004. Ectopic expression of a *Haemonchus contortus* GATA transcription factor in *Caenorhabditis elegans* reveals conserved function in spite of extensive sequence divergence. *Molecular and Biochemical Parasitology,* 133 (2), 241-253.

Dhanasekaran, N. and Dermott, J.M., 1996. Signaling by the G_{12} class of G proteins. *Cellular Signalling,* 8 (4), 235-245.

Dusenbery, D.B., 1974. Analysis of chemotaxis in the nematode *Caenorhabditis elegans* by counter current separation. *Journal of Experimental Zoology,* 188 (1), 41-47.

Fujiwara, M., Sengupta, P. and McIntire, S.L., 2002. Regulation of body size and behavioral state of *C. elegans* by sensory perception and the EGL-4 cGMP-dependent protein kinase. *Neuron,* 36 (6), 1091-1102.

Gommers, F.J. and Bakker, J., 1988. Physiological diseases induced by plant responses or products. *In:* Poinar, G.O. and Jansson, H.B. eds. *Diseases of nematodes*, Vol. 1. CRC Press, Boca Raton, 3-22.

Graziano, M.P., Casey, P.J. and Gilman, A.G., 1987. Expression of complementary DNA species for G proteins in *Escherichia coli* two forms of Gs alpha stimulate adenylate cyclase. *Journal of Biological Chemistry,* 262 (23), 11375-11381.

Hamm, H.E., 1998. The many faces of G protein signaling. *Journal of Biological Chemistry,* 273 (2), 669-672.

Hashmi, S., Hashmi, G., Glazer, I., et al. 1998. Thermal response of *Heterorhabditis bacteriophora* transformed with the *Caenorhabditis elegans hsp70*-encoding gene. *Journal of Experimental Zoology,* 281 (3), 164-170.

Hilliard, M.A., Bargmann, C.I. and Bazzicalupo, P., 2002. *C. elegans* responds to chemical repellents by integrating sensory inputs from the head and the tail. *Current Biology,* 12 (9), 730-734.

Hirotsu, T., Saeki, S., Yamamoto, M., et al. 2000. The Ras-MAPK pathway is important for olfaction in *Caenorhabditis elegans*. *Nature,* 404 (6775), 289-293.

Hoschitz, M., Buchholz, T.G. and Ott, J.A., 1999. *Leptonemella juliae* sp.n. and *Leptonemella vestari* sp.n. (Stilbonematinae), two new free-living marine nematodes from a subtidal sand bottom. *Annalen des Naturhistorischen Museums in Wien. Serie B. Botanik und Zoologie* (101), 423-435.

Jansen, G., Thijssen, K.L., Werner, P., et al. 1999. The complete family of genes encoding G proteins of *Caenorhabditis elegans*. *Nature Genetics,* 21 (4), 414-419.

Krieger, J. and Breer, H., 1999. Olfactory reception in invertebrates. *Science,* 286 (5440), 720-723.

Kwa, M.S.G., Veenstra, J.G., Dijk, M.V., et al. 1995. [beta]-Tubulin genes from the parasitic nematode *Haemonchus contortus* modulate drug resistance in *Caenorhabditis elegans*. *Journal of Molecular Biology,* 246 (4), 500-510.

L'Etoile, N.D. and Bargmann, C.I., 2000. Olfaction and odor discrimination are mediated by the *C. elegans* guanylyl cyclase ODR-1. *Neuron,* 25 (3), 575-586.

L'Etoile, N.D., Coburn, C.M., Eastham, J., et al. 2002. The cyclic GMP-dependent protein kinase EGL-4 regulates olfactory adaptation in *C. elegans*. *Neuron,* 36 (6), 1079-1089.

Linder, M.E., Ewald, D.A., Miller, R.J., et al. 1990. Purification and characterization of Go alpha and three types of Gi alpha after expression in *Escherichia coli*. *Journal of Biological Chemistry,* 265 (14), 8243-8251.

Masamune, T., Anetai, M., Takasugi, M., et al. 1982. Isolation of a natural hatching stimulus, glycinoeclepin A, for the soybean cyst nematode. *Nature,* 297 (5866), 495-496.

Mendel, J.E., Korswagen, H.C., Liu, K.S., et al. 1995. Participation of the protein Go in multiple aspects of behavior in *C. elegans*. *Science,* 267 (5204), 1652-1655.

Miller, K.G., Emerson, M.D. and Rand, J.B., 1999. Go[alpha] and diacylglycerol kinase negatively regulate the Gq[alpha] pathway in *C. elegans*. *Neuron,* 24 (2), 323-333.

Milligan, G., Mullaney, I. and McCallum, J.F., 1993. Distribution and relative levels of expression of the phosphoinositidase-C-linked G-proteins G_q alpha and G_{11} alpha: absence of G_{11} alpha in human platelets and haemopoietically derived cell lines. *Biochimica et Biophysica Acta,* 1179 (2), 208-212.

Mongan, N.P., Baylis, H.A., Adcock, C., et al. 1998. An extensive and diverse gene family of nicotinic acetylcholine receptor alpha subunits in *Caenorhabditis elegans*. *Receptors Channels,* 6 (3), 213-228.

O'Halloran, D.M. and Burnell, A.M., 2003. An investigation of chemotaxis in the insect parasitic nematode *Heterorhabditis bacteriophora*. *Parasitology,* 127 (4), 375-385.

Parks, S. and Wieschaus, E., 1991. The *Drosophila* gastrulation gene *concertina* encodes a G alpha-like protein. *Cell,* 64 (2), 447-458.

Pline, M. and Dusenbery, D.B., 1987. Responses of plant-parasitic nematode *Meloidogyne incognita* to carbon dioxide determined by video camera-computer tracking. *Journal of Chemical Ecology,* 13 (4), 873-888.

Poinar, G.O., 1983. *The natural history of nematodes*. Prentice-Hall, Englewood Cliffs.

Prasad, B.C. and Reed, R.R., 1999. Chemosensation: molecular mechanisms in worms and mammals. *Trends in Genetics,* 15 (4), 150-153.

Rasmann, S., Kollner, T.G., Degenhardt, J., et al. 2005. Recruitment of entomopathogenic nematodes by insect-damaged maize roots. *Nature,* 434 (7034), 732-737.
Riddle, D.L., Swanson, M.M. and Albert, P.S., 1981. Interacting genes in nematode dauer larva formation. *Nature,* 290 (5808), 668-671.
Riemann, F. and Schrage, M., 1988. Carbon dioxide as an attractant for the free-living marine nematode *Adoncholaimus thalassophygas*. *Marine Biology,* 98 (1), 81-85.
Riga, E., Perry, R.N., Barrett, J., et al. 1996. Electrophysiological responses of males of the potato cyst nematodes, *Globodera rostochiensis* and *G. pallida*, to their sex pheromones. *Parasitology,* 112 (2), 239-246.
Robertson, H.M., 1998. Two large families of chemoreceptor genes in the nematodes *Caenorhabditis elegans* and *Caenorhabditis briggsae* reveal extensive gene duplication, diversification, movement, and intron loss. *Genome Research,* 8 (5), 449-463.
Robertson, H.M., 2000. The large *srh* family of chemoreceptor genes in *Caenorhabditis* nematodes reveals processes of genome evolution involving large duplications and deletions and intron gains and losses. *Genome Research,* 10 (2), 192-203.
Robertson, H.M., 2001. Updating the *str* and *srj* (*stl*) families of chemoreceptors in *Caenorhabditis* nematodes reveals frequent gene movement within and between chromosomes. *Chemical Senses,* 26 (2), 151-159.
Robinson, A.F., 1995. Optimal release rates for attracting *Meloidogyne incognita, Rotylenchulus reniformis*, and other nematodes to carbon dioxide in sand. *Journal of Nematology,* 27 (1), 42-50.
Ross, E.M. and Wilkie, T.M., 2000. GTPase-activating proteins for heterotrimeric G proteins: regulators of G protein signaling (RGS) and RGS-like proteins. *Annual Review Biochemistry,* 69, 795-827.
Ruvkun, G. and Hobert, O., 1998. The taxonomy of developmental control in *Caenorhabditis elegans*. *Science,* 282 (5396), 2033-2041.
Schöller, C., Molin, S. and Wilkins, K., 1997. Volatile metabolites from some gram-negative bacteria. *Chemosphere,* 35 (7), 1487-1495.
Sengupta, P., Chou, J.H. and Bargmann, C.I., 1996. odr-10 encodes a seven transmembrane domain olfactory receptor required for responses to the odorant diacetyl. *Cell,* 84 (6), 899-909.
Simon, M.I., Strathmann, M.P. and Gautam, N., 1991. Diversity of G proteins in signal transduction. *Science,* 252 (5007), 802-808.
Snow, B.E., Hall, R.A., Krumins, A.M., et al. 1998. GTPase activating specificity of RGS12 and binding specificity of an alternatively spliced PDZ (PSD-95/Dlg/ZO-1) domain. *Journal of Biological Chemistry,* 273 (28), 11749-17755.
Stein, L.D., Bao, Z.R., Blasiar, D., et al. 2003. The genome sequence of *Caenorhabditis briggsae*: a platform for comparative genomics. *PLoS Biology,* 1 (2), 166-192.
Strathmann, M.P. and Simon, M.I., 1991. G[alpha]12 and G[alpha]13 subunits define a fourth class of G protein alpha subunits. *Proceedings of the National Academy of Sciences of the United States of America,* 88 (13), 5582-5586.
Tesmer, J.J., Sunahara, R.K., Gilman, A.G., et al. 1997. Crystal structure of the catalytic domains of adenylyl cyclase in a complex with G_{salpha}.$GTP_{gamma}S$. *Science,* 278 (5345), 1907-1916.
Thomas, J.H., Birnby, D.A. and Vowels, J.J., 1993. Evidence for parallel processing of sensory information controlling dauer formation in *Caenorhabditis elegans*. *Genetics,* 134 (4), 1105-1117.
Thomas, J.H., Kelley, J.L., Robertson, H.M., et al. 2005. Adaptive evolution in the SRZ chemoreceptor families of *Caenorhabditis elegans* and *Caenorhabditis briggsae*. *Proceedings of the National Academy of Sciences of the United States of America,* 102 (12), 4476-4481.
Treinin, M. and Chalfie, M., 1995. A mutated acetylcholine receptor subunit causes neuronal degeneration in *C. elegans*. *Neuron,* 14 (4), 871-877.
Troemel, E.R., Chou, J.H., Dwyer, N.D., et al. 1995. Divergent seven transmembrane receptors are candidate chemosensory receptors in *C. elegans*. *Cell,* 83 (2), 207-218.
Urwin, P.E., Lilley, C.J. and Atkinson, H.J., 2002. Ingestion of double-stranded RNA by preparasitic juvenile cyst nematodes leads to RNA interference. *Molecular Plant Microbe Interactions,* 15 (8), 747-752.
Van der Voorn, L., Gebbink, M., Plasterk, R.H.A., et al. 1990. Characterization of a G-protein [beta]-subunit gene from the nematode *Caenorhabditis elegans*. *Journal of Molecular Biology,* 213 (1), 17- 26.

Ward, S., 1973. Chemotaxis by the nematode *Caenorhabditis elegans*: identification of attractants and analysis of the response by use of mutants. *Proceedings of the National Academy of Sciences of the United States of America,* 70 (3), 817-821.

Watson, N., Linder, M.E., Druey, K.M., et al. 1996. RGS family members: GTPase-activating proteins for heterotrimeric G-protein alpha-subunits. *Nature,* 383 (6596), 172-175.

White, J.G., Albertson, D.G. and Anness, M.A.R., 1978. Connectivity changes in a class of motoneurone during the development of a nematode. *Nature,* 271, 764-766.

White, J.G., Southgate, E. and Thompson, J.N., 1991. On the nature of undead cells in the nematode *Caenorhabditis elegans. Philosophical Transactions of the Royal Society of London,* 331, 263-271.

White, J.G., Southgate, E., Thomson, J.N., et al. 1986. The structure of the nervous system of the nematode *Caenorhabditis elegans. Philosophical Transactions of the Royal Society of London. Series B. Biological Sciences,* 314 (1165), 1-340.

Zechman, J.M. and Labows, J.N., 1985. Volatiles of *Pseudomonas aeruginosa* and related species by automated headspace concentration-gas chromatography. *Canadian Journal of Microbiology,* 31 (3), 232-237.

Zhang, Y., Chou, J.H., Bradley, J., et al. 1997. The *Caenorhabditis elegans* seven-transmembrane protein ODR-10 functions as an odorant receptor in mammalian cells. *Proceedings of the National Academy of Sciences of the United States of America,* 94 (22), 12162-12167.

CHAPTER 7

VARIATION IN LEARNING OF HERBIVORY-INDUCED PLANT ODOURS BY PARASITIC WASPS

From brain to behaviour

HANS M. SMID

Laboratory of Entomology, Wageningen University, Binnenhaven 7, 6709 PD Wageningen, The Netherlands. E-mail: hansm.smid@wur.nl

Abstract. Two closely related parasitic wasp species, *Cotesia glomerata* and *Cotesia rubecula*, lay their eggs in first-instar caterpillars of *Pieris brassicae* and/or *Pieris rapae* hosts. They find their hosts by responding to secondary plant metabolites, induced by herbivory. Both wasp species have an innate preference for the odours of infested cabbage, common host plants of these *Pieris* caterpillars, but they can also learn to respond to the odours of other host plants, after they have found suitable host caterpillars on that plant. This experience results in an association of the odours of that plant with the presence of suitable hosts. The two wasp species differ profoundly in olfactory learning; *C. glomerata* instantly changes its innate preference for cabbage odours towards the odours of another plant after a single experience, whereas *C. rubecula* never changes its innate preference for cabbage odours. Both wasps show an increase in flight response to a previously unattractive host plant after a single oviposition experience on that plant, but this memory wanes in *C. rubecula* after a day, and remains unchanged for at least 5 days in *C. glomerata*.

In this paper, ultimate factors are discussed that may have contributed to the evolution of the observed differences in learning in these two wasp species. Furthermore, hypotheses on the possible neural mechanisms and genes underlying these differences are given, based on current knowledge on the cellular mechanisms of learning as determined for genetic and neurobiological model species like the fruit fly *Drosophila melanogaster* and the honeybee *Apis mellifera*.
Keywords: learning; memory; olfaction; parasitoid; *Cotesia*; synaptic plasticity; octopamine; CREB; conditioning

INSECTS AND LEARNING

Many people have the idea that insects are little programmed machines, designed to perform a set of simple behaviours in a fixed way, and that they are in no way functionally comparable to higher animals. Current research has shown this idea to be entirely wrong (Collett and Collett 2002; Giurfa 2003; Watanabe et al. 2003). It may feel uncomfortable to man, but inside the head of, e.g., a tiny fruitfly exists a brain of a mere halve millimetre, housing some 200,000 neurons that function in a way that is not different from the 100 billion neurons in our human brains (Figure 1). The networks formed by the fly's neurons result in a functional brain with

M. Dicke and W. Takken (eds.), Chemical Ecology: From Gene to Ecosystem, 89-103.

remarkable capacities, including the ability to learn. It is obvious that the fly's cognitive possibilities are limited, but it is well equipped to respond in a flexible way to its environment, and to gain from its previous experiences. Thus, an experienced insect can display a dramatically different behaviour compared to a naive insect through learning, and this learning effect can last for the rest of the insect's life.

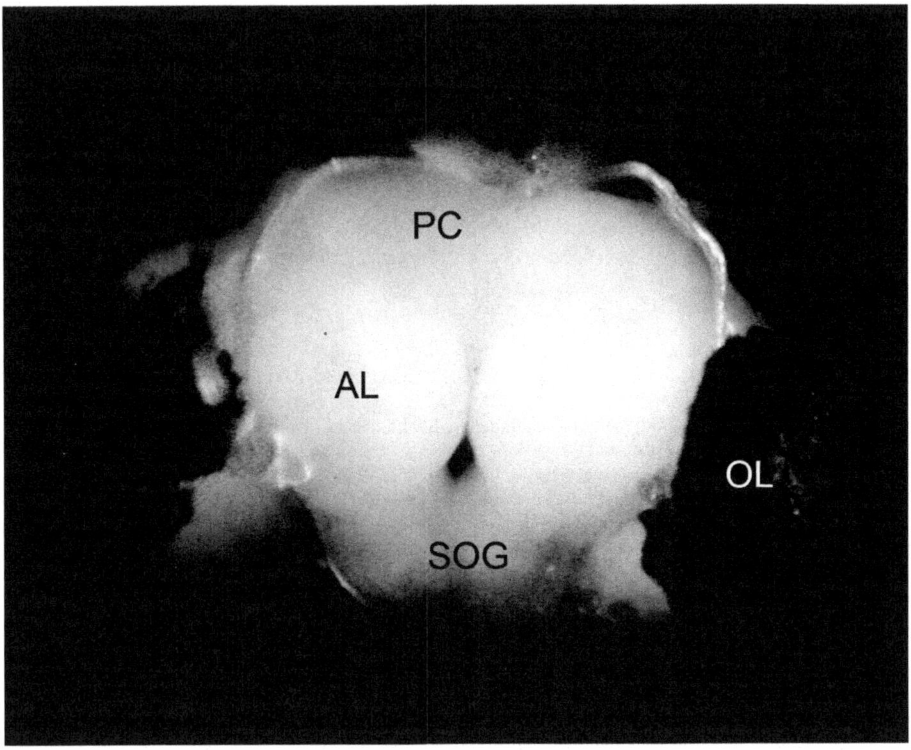

Figure 1. *Brain of a parasitic wasp, Cotesia glomerata. The size of this brain is approximately 750 μm width. AL, antennal lobe; OL, optic lobe; SOG, suboesophageal ganglion; OC, ocellus; PC, protocerebrum*

Insects are well equipped for associative learning. They can learn quickly to search for items by responding to a cue that has previously been rewarded, or they can avoid cues that were sensed within a negative experience. Within the context of the theme of this volume, I will focus on olfactory learning in parasitic wasp species attacking the larval stage of cabbage butterflies (*Pieris* spp.). Having said that, insects are by no means limited to olfactory learning alone.

PARASITIC WASPS AND ASSOCIATIVE LEARNING

Parasitic wasps lay their eggs in or on their insect hosts. There are wasps that parasitize eggs, larvae, pupae or adults. One example is the species *Cotesia glomerata* (Figure 2). This wasp species lays its eggs in young larvae of Pierid butterflies, such as the large and small cabbage white, *Pieris brassicae* and *P. rapae*, resp. The eggs of the wasp hatch inside the body of the caterpillar and both the caterpillar and wasp larvae develop until they reach the final larval instar. At that point, the wasp larvae eat their way out through the cuticle of the caterpillar, spin a

Figure 2. *Cotesia glomerata ovipositing in a Pieris brassicae caterpillar*

cocoon and moult to the pupal stage, leaving the dying caterpillar behind. The adult wasps that emerge from the cocoons have three different foraging tasks: (1) to find a mate, (2) to find food (nectar or honeydew) and (3) to find host larvae to lay their eggs. The latter foraging task, which is obviously only relevant for females, will be the focus of this paper. The tiny young host caterpillars take care not to spread attractants, they are well camouflaged and do not emit odorants themselves that can be perceived by the wasps from a distance. However, feeding by the caterpillars on their host plants induces the emission of volatiles from their food plant, and these are highly attractive to the wasp. In the case of *Cotesia glomerata*, the odour of cabbage damaged by feeding *P. brassicae* larvae is highly attractive (Geervliet et al. 1994). The response to the odour of damaged cabbage plants is high in naive wasps, and it is not necessary for the wasp to learn to recognize this odour (Geervliet et al. 1996). However, wasps can learn to associate odours from another plant species to the presence of suitable hosts, if they have an oviposition experience on that plant. In this way they can learn to change their foraging behaviour; after the experience on a different plant species, they will specifically search for that plant species (Geervliet et al. 1998). They have learned to associate the odours of that plant with the presence of suitable host caterpillars.

DIFFERENT FORMS OF LEARNING

To understand learning behaviour it is essential to discriminate between the different forms of learning (Rescorla 1988; Krasne and Glanzman 1995). Most important, there is associative learning but there are also simpler, non-associative forms. Sensitization and habituation are examples of non-associative learning. During habituation, an animal is repeatedly stimulated by a stimulus which results in a gradually lower response to that stimulus. Habituation can last minutes to hours, and if many repetitions of the stimulus are given, even much longer. This form of non-associative learning is very important for an animal, because it enables the animal to learn to ignore unimportant stimuli. Habituation is the result of active suppression of the response. It is a process that is different (though often difficult to separate) from sensory adaptation or muscle fatigue. This can be demonstrated when the habituated stimulus is given after a noxious stimulus (e.g., a shock). The response to the habituated stimulus is then completely recovered; this process is called dis-habituation (Corfas and Dudai 1989). This phenomenon ensures that an animal can adequately respond to a previously habituated stimulus after an important change in the situation, as perceived by, e.g., a noxious stimulus.

Sensitization occurs when an animal is stimulated by a significant stimulus, such as a shock, a reward, a loud noise or strong odour. This form of non-associative learning is non-specific, and it is characterized by a general increase in response to other stimuli. This effect can, like habituation, last from minutes to hours or even weeks. Sensitization enables an animal to respond better to stimuli when confronted with a significant change in the situation, as signalled by the stimulus that induces sensitization.

Associative learning is different from habituation and sensitization in that it can only occur if two stimuli are presented to the animal; a neutral stimulus, immediately followed by a meaningful, reinforcing stimulus, which can be a reward or a punishment. The animal learns to associate the neutral stimulus with the reinforcing stimulus. This form of learning is called classical conditioning or Pavlovian conditioning, after the famous researcher Pavlov (1927). He trained his dog to respond to a sound, by giving it food as a reward each time it heard the sound. After several pairings of the sound and the reward, the dog started salivating when it heard the sound only. The neutral stimulus that becomes associated with the reward is called the conditioned stimulus (CS), the reinforcer (the reward or punishment) is called the unconditioned stimulus (US). The dog shows the unconditioned response (UR, salivation) to the US. Only after conditioning, it shows the conditioned response (CR, also salivation) to the CS. Only if the CS is directly followed by the US, which is called forward pairing, associative learning will occur. Backward pairing (US followed by CS) does not result in associative learning, but sensitization can occur, resulting in a temporary increase of the response to the CS.

Another form of associative learning is operant conditioning (Thorndike 1901; Skinner 1938). Here, a CS triggers a specific behavioural response, which is followed by the reinforcer only when this appropriate behavioural response is performed. Thus, it is the behavioural response to the CS which is reinforced by the US; an association is formed between the response to the CS and the reinforcer, rather than between the CS and the US as in classical conditioning. However, elements of classical conditioning are present in operant conditioning as well, and it is sometimes difficult to determine whether a form of learning is purely classical conditioning or operant conditioning (see for a comprehensive overview of both forms of conditioning: Lutz (1994)).

PARASITIC-WASP LEARNING AS CLASSICAL CONDITIONING

How does the theory of associative learning translate into the example of parasitic-wasp olfactory learning? The wasp first smells the odour of a plant on which host caterpillars are feeding, and responds to it by landing on a leaf, where it is rewarded with the presence of suitable host caterpillars. Thus, this is a clear form of operant conditioning. However, in the laboratory set-up, the flight response is usually not incorporated in the learning trial, for reasons of standardization and convenience. The wasp is placed directly on the leaf with caterpillars and stimulated by the taste of host-derived substances such as faeces (frass). The mere perception of host traces on the leaf without any behavioural response is sufficient to learn to recognize the odours of the leaf (Geervliet et al. 1998), which is in line with the CS-US contingency of classical conditioning. The odour of the leaf is the CS and the taste of host-derived substances is the US in this case. Thus, an association between the plant odours and the presence of suitable host caterpillars is already made before oviposition has taken place. However, the increased response to the plant odours lasts longer when the perception of host traces is followed by oviposition (Takasu and Lewis 2003). The test for memory formation is done in a wind tunnel set-up, where it is given a choice to fly towards the naively preferred plant or to the experienced plant. This is actually an operant context; the wasp has learned to adapt its flight response in a classical conditioning learning paradigm. Such a transfer of information has been demonstrated also in another learning paradigm, the proboscis extension reflex of the honeybee (Bitterman et al. 1983). Here, a harnessed honeybee learns to associate an odour with the reward, an application of a droplet of sugar water on its antennae in a purely classical conditioning set-up. Before the learning trial, the bee responds to the US (sugar water) by extension of its proboscis (UR), and after the learning trial, the bee responds to the odour (CS) by extension of the proboscis (CR). Also in this case, the trained bee shows that this experience changes searching behaviour under free flight conditions (Sandoz et al. 2000), thus in a operant context.

DURATION OF MEMORY

Learning results in the formation of memory. Like in higher animals, insects have different forms of memory, ranging from short-term to long-term memories. A number of factors influence what kind of memory is formed; the strengths of both US and CS, the number of repetitions of US-CS pairings, the time interval between the repetitions (the inter-trial interval) and the time interval between the US and CS (the inter-stimulus interval) (Menzel 1999; Menzel et al. 2001). In general, the longer the inter-stimulus interval, the weaker the association between the stimuli. A single US-CS pairing usually results in memory that lasts no more than hours or one day. If repeated US-CS pairings occur, long-term memory can be formed, but only if there is some time in between each learning trial. This is called spaced learning trials, in contrast to massed learning, which means that the learning trials are in rapid succession following each other. Massed learning is not effective to induce the formation of long-term memory, whereas an identical number of spaced learning trials is (Menzel et al. 2001).

In the case of a parasitic wasp that encounters a gregarious host (e.g., several caterpillars feeding together on a single leaf), the rapid sequence of ovipositions should probably be considered a mass learning experience, although the breaks that wasps often take after several ovipositions before they resume their attacks may interfere with this conclusion. Only after the wasp leaves the plant and lands on another plant of the same species and encounters host frass, an additional conditioning trial occurs that matches the criteria for spaced learning trials.

The mechanism that a long-term, stable memory is only formed after several experiences makes sense. This way, only relevant, reliable information is stored. A single experience results in a less stable memory form that wanes if it is not reinforced by additional experiences. This mechanism of memory formation serves as a filter that ensures that only important and reliable information is stored in long-term memory. It is not efficient to learn too fast, because this way an animal will easily store the wrong information.

DIFFERENCES IN LEARNING BETWEEN CLOSELY RELATED SPECIES

Closely related species, for example of parasitic wasps, may display large differences in learning (Poolman-Simons et al. 1992; Potting et al. 1997; Geervliet et al. 1998). Such closely related species are ideal subjects for a comparative approach.

The wasp, *C. glomerata* learns to change its innate preference for cabbage plant odours towards the odours of another plant species, *Nasturtium,* which is an alternative host plant of *P. brassicae.* One day after the learning trial (an oviposition experience on a *Nasturtium* leaf) the wasps were released in a wind tunnel where they could choose between a leaf from a cabbage plant and one from a *Nasturtium* plant, both infested with the same number of host caterpillars. The wasps did no longer prefer cabbage but landed on the *Nasturtium* leaves (Geervliet et al. 1998). However, when this experiment was performed with *C. rubecula,* which is a closely related species (Michel-Salzat and Whitfield 2004), the preference shift did not occur. Even after several experiences, the wasps continued to choose the cabbage odours for which they had an innate preference. In a subsequent study, Bleeker et al. (in press-a) investigated this phenomenon in a different way. They did not measure a change in preference using a wind-tunnel set-up with a choice situation (a dual-choice set-up), but measured the flight response to the learned plant odour with a single-choice set-up (also called a no-choice set-up), using controls that distinguished sensitization from associative learning. This way, the increase in response to the *Nasturtium* odours can be determined irrespective of the strength of the innate preference to cabbage odours. The wasps were released in the wind tunnel and the number of wasps that landed on the *Nasturtium* leaves was determined before and at different time intervals after the oviposition experience. Two remarkable results were obtained in this study. First, *C. glomerata* formed a long-lasting memory of at least 5 days after a single oviposition experience, which is in contrast to the notion that several spaced learning trials are necessary for long lasting memory. Second, it was shown that *C. rubecula* learned to respond to the *Nasturtium* odours by associative learning, even though this memory waned gradually after one day. Thus, *C. rubecula* does learn to associate the odours of *Nasturtium* to the presence of hosts (albeit that this memory lasts much shorter than in *C. glomerata*), but does not change its innate preference for cabbage odours. These two wasp species are closely related, have very similar morphology, including brain morphology (Smid et al. 2003), olfactory sensilla morphology (Bleeker et al. 2004) and olfactory receptive range (Smid et al. 2002). The difference in learning may be an adaptation to the specific differences in host-finding behaviour between the two wasp species.

ULTIMATE FACTORS THAT MAY BE CORRELATED TO VARIATION IN LEARNING

There are a number of ultimate factors commonly associated with variation in learning (see for reviews Shettleworth 1993; Turlings et al. 1993; Vet et al. 1995). What factors drive the evolution of slow learners (i.e., species that need many repeats of experiences to adapt their behaviour), and what factors drive towards fast learning (i.e., species that adapt their behaviour upon a single experience)? What are the limits to the amount of information that can be stored in the tiny brains of parasitic wasps? It is important to realize that learning and memory are costly processes at different levels. There is the energy cost of learning and of the formation and maintenance of memory (Dukas 1999; Mery and Kawecki 2003; 2005). In addition to the energetic costs of memory there is also the ecological cost. Learning trials take time and may constitute a risk (e.g., predation) compared to innate behaviour (see below), and the information that is learned may be wrong, leading to maladaptive behaviour with strong fitness penalties. However, these ecological costs to learning can be different between species (Dukas 1998b).

First, the life span of the animal is important. It is obviously not useful to spend much time on learning for an insect that lives for only one day. Also the number of foraging decisions that an insect makes, determines that a certain subset of these decisions can be spent to learn to optimize the remaining part of the foraging decisions (Roitberg et al. 1993; Dukas and Kamil 2001). Learning takes time, hence there must be a certain optimum of learning trials to spend on learning, and if time is relatively costly, like it is for a short-lived insect, this optimum will be driven towards fewer learning trials.

Another important factor is the reliability of the association that is learned. If the associations that are learned are very variable, more learning trials are necessary before the animal should adapt its behaviour to a supposed relevant new situation.

The strengths of fitness penalties that come with wrong associations are important factors that drive evolution of slower learning. If the information is less variable, fewer learning trials are needed to change behaviour. Thus, the variability of the environment may influence the learning speed of the insect; an insect species living in a less variable environment may evolve towards a fast-learning species, whereas an insect living in a highly variable environment, where there is little predictive value from a single experience, may evolve towards a slow-learning species.

Stephens (1993) distinguished within-generation variability and between-generation variability. In this model, learning is favoured when between-generation predictability is low but within-generation predictability is high. If between-generation predictability is high, innate behaviour is expected to be favoured over learning. Parasitic wasps have a strong innate response to, e.g., the taste of sugars or the taste of host-derived substances, as these substances are invariable between generations. However, the response to plant odours on which the parasitoid's host may occur depends on the predictive value of that odour to the wasp. When the host occurs only on that plant species, the predictive value is high and a strong innate response is favourable (Vet et al. 1990; Vet and Dicke 1992). If the host occurs on several plant species, the innate responses to those plant species are expected to be intermediate, but become stronger after repeated experiences. If the host occurs on several other plant species that are also available in the area, the predictive value is low. The wasp needs to respond to all odours of potential host plants on which its host may be present, and it has to divide its attention over a wide range of plant species. This is disadvantageous for two reasons. First, because herbivory-induced plant odour blends are difficult to detect against a background of non-relevant plant odours, detection becomes less efficient when the wasp has to divide its attention to several different potentially relevant stimuli (the problem of limited attention, Bernays and Wcislo 1994; Bernays 1996; Dukas 1998a; 1998b; Dukas and Kamil 2001). Second, the wasp needs to spend time to visit several host plants that will not be rewarding. Specialization is thought to be an adaptation to this problem. Preference learning can be seen as a way to achieve temporal specialization (Dukas 1998b). Thus, learning is a trait that may be tightly linked to the level of specialization of both the wasp and the host, and especially generalist wasps that parasitize on generalist hosts may benefit from learning by gaining from the advantages of specialization.

In conclusion, there are several factors that may influence learning of a parasitic wasp. How can these ideas help us to understand the difference in learning between *C. glomerata* and *C. rubecula?*

In the case of *C. rubecula*, the wasp remembers the odours of a plant after an experience for a short term, but does not change its preference. Apparently, the cabbage odours remain the most reliable indicators for the presence of hosts. This may be an adaptation to the oviposition behaviour of its host, which is the small cabbage white, *Pieris rapae*. This butterfly lays only a single egg on a plant and does this in a rather unpredictable way (Root and Kareiva 1984; Davies and Gilbert 1985), possibly to avoid parasitization. Thus, the predictive value of a single oviposition experience for *C. rubecula* is low. Apparently, the trait of learning in the brain of this wasp species is adapted to the oviposition behaviour of its host species.

Moreover, *C. rubecula* is a solitary species that lays only one egg into a host, and has to find a large number of hosts on a large number of plants. Thus it makes a large number of foraging decisions. This may also contribute to the slow learning speed.

For *C. glomerata*, the situation is profoundly different. This species is more of a generalist than *C. rubecula*, but our population, which was collected in The Netherlands, strongly prefers the large cabbage white, *Pieris brassicae*. This butterfly species lays clusters of eggs. The caterpillars, after hatching, completely destroy the plant on which they are feeding and subsequently have to migrate to neighbouring plants. Due to their induced dietary specialization, they need to migrate to the same plant species as the one on which they initiated feeding, and therefore the butterfly has to lay its eggs on dense stands of plants of the same species (Le Masurier 1994). Such dense stands are likely to attract more ovipositing butterflies than single plants, and that may be a reason that a single oviposition experience of *C. glomerata* on *P. brassicae* is reliable enough to induce long-term memory formation. Moreover, *C. glomerata* is a gregarious wasp that lays several eggs into a single caterpillar. This means that it can oviposit half of its lifetime fecundity into a single clutch of caterpillars. Thus it needs only a few foraging decisions (see Roitberg et al. 1993), and it may well be the optimal strategy to learn to keep searching on the dense stand of the same plant species because chances are high to discover another rich source of oviposition opportunities.

NEURAL BACKGROUNDS: WHAT HAPPENS IN THE BRAIN DURING LEARNING AND MEMORY FORMATION?

In order to understand how evolution shapes learning, it is crucial to identify the neural mechanisms that are underlying these differences in learning. Which genes and which neurons are involved, and how do they encode for the differences in learning? Only if this information becomes available, will it be feasible to study variations in those genes and neurons in a large number of different species and predict variation in learning ability and correlations with ultimate factors. I will first focus on the level of small neural networks to explain what happens during classical conditioning in the brain of an insect, and then descend to the molecular level to focus on some genes that play a key role in learning.

When an association is made between a CS and a US, it is obviously necessary that the neural responses to these two stimuli are somehow brought together in the brain. Either a reward or a punishment can serve as reinforcement in olfactory conditioning that stimulates formation of odour memory, so that an association is made between the odour and the reward or punishment. How this mechanism works in the case of reward learning was described for the honeybee (Hammer 1993; 1997). The honeybee is a well-known model animal for neurobiological research on classical conditioning. Much research is done using proboscis extension reflex (PER) conditioning. Honeybees extend their mouthparts (proboscis) when stimulated by sugar solution on the taste sensilla on the antenna and mouthparts. This reflex can be conditioned when the sugar stimulus is preceded by an odour stimulus. The bee learns that the odour predicts the sugar reward, and subsequently extends its proboscis when the odour stimulus is presented alone. Hammer studied the electrical properties of an intriguing neuron that innervates the entire olfactory pathway in the honeybee and releases the neuromodulator octopamine, a substance known to mediate the reward in classical conditioning in insects (Hammer and Menzel 1998; Schwaerzel et al. 2003). This neuron was among a group of neurons lying ventrally along the mid axis of the brain called ventral unpaired median neurons or VUM neurons. The VUM neuron studied by Hammer sends its arborizations bilaterally into the entire olfactory pathway. Hammer succeeded in making electrical recordings of this VUM neuron in the honeybee's brain while performing PER conditioning, and found that this neuron responded strongly when the honeybee was stimulated with the sugar reward. When he applied the odour to the antenna, and subsequently stimulated the VUM neuron artificially (without sugar application but by electrical stimulation of the cell body), he could achieve PER conditioning to the same extent as with a sugar reward. Thus, the sugar reward could be entirely substituted by stimulation of a single neuron. This very simple network gives us a clear idea how learning acts at the neural level and how a reward-sensitive neuron plays a key role in this process. This VUM neuron belongs to a group of other VUM neurons that also express the neuromodulator octopamine (Kreissl et al. 1994), but with different arborization pathways, projecting, e.g., towards the optic lobes or into the antenna (Schröter 2002). Thus, there are most likely more reward-sensitive VUM neurons involved.

Octopaminergic VUM neurons are present in parasitic wasps as well (Smid et al. 2003; Bleeker et al. in press-b, Figure 3), and are candidate neurons that may encode for differences in learning observed in species like *C. glomerata* and *C. rubecula*. For instance, the strength of the response to a reward may be different, or the density of their arborizations into the olfactory pathway, leading to differences in the amount of octopamine released in the olfactory pathway upon a reward stimulus. Another explanation for the observed difference in learning may lie in the sensitivity to the octopamine signal in the olfactory pathway. To understand this, it is necessary to focus on the molecular level of learning.

MOLECULAR BIOLOGY: CANDIDATE GENES ENCODING LEARNING DIFFERENCES

The location where memory is stored at the level of single neurons is the synapse. The cellular equivalent of memory is called synaptic plasticity (Pittenger and Kandel 2003). The properties of a synapse can change after previous activity, and this is the way how the neuron 'remembers' its previous activity. For instance, transmission of a signal may be facilitated by increasing the amount of synaptic vesicles that are released upon electric stimulation, and hence increasing the post-synaptic response levels. The duration of these and other processes that underlie synaptic plasticity, is corresponding with short- or medium-term memory. This is called long-term

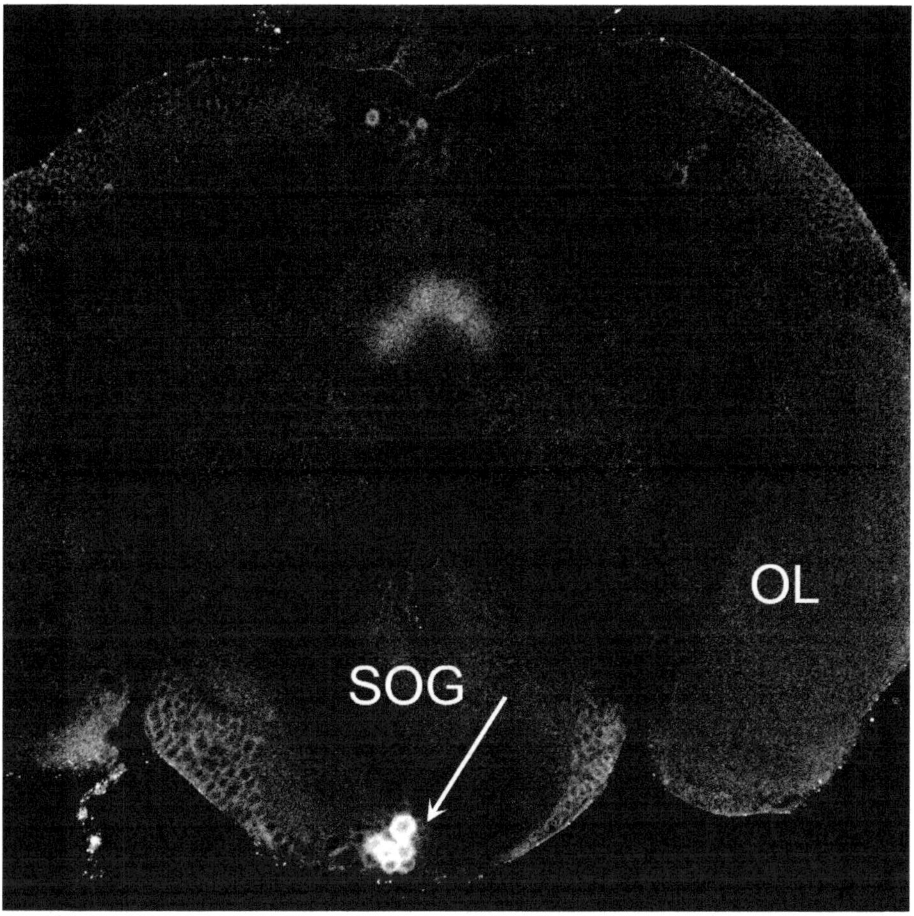

Figure 3. Confocal section of a brain of C. rubecula, showing cell bodies of a group of VUM neurons (arrow). These neurons were visualized using a fluorescently labelled antibody against octopamine. OL, optic lobe; SOG, suboesophagal ganglion

potentiation, and the opposite process, long-term depression, can also occur (Huber et al. 2000). There is also an equivalent form of long-term memory, called late long-term potentiation, when synaptic transmission becomes facilitated by the growth of new synaptic contacts. This way, the number of synaptic connections between two neurons is increased, and the increased synaptic strength that is formed by this process results in a more stable form of synaptic potentiation. This mechanism requires gene transcription and the production of new proteins (Nguyen et al. 1994). The learning-induced changes in synapse properties, either long-term or short-term, can occur throughout the brain in various neuropiles that are involved in, e.g., a learned behaviour, rather than at a specific region in the brain dedicated to memory storage. Hence the term 'memory trace' is used to refer to the changes in neural elements were memory is stored.

The cellular pathways involved in synaptic plasticity are remarkably conserved within the animal kingdom, and it is now well-known that the cAMP – protein kinase A (PKA) signalling pathway plays a central role in species varying from nematodes, snails and insects to mammals (Silva et al. 1998; Eisenhardt in press). Single learning experiences induce a limited elevation of cAMP levels, which activates PKA, acting locally at the site of the synapse. Repeated learning experiences result in the activation of larger amounts of activated PKA, which translocates towards the nucleus where it activates a transcription factor. This transcription factor, called cAMP-responsive element binding protein (CREB) causes the expression of genes that are necessary to produce the proteins required for the formation of stable, long-term memory. There are several different isoforms of CREB resulting from alternative splicing that are different in the way they respond to PKA. Some isoforms are lacking parts of the amino-acid sequences that allow binding and activation by PKA, and therefore inhibit long-term memory formation. The different CREB isoforms represent memory suppressor as well as memory enhancer isoforms, and it is thought that the balance of tissue-specific expression of these isoforms determines the sensitivity of a neuron for the cAMP signal, and thus to the US (Yin et al. 1994; 1995a; Yin and Tully 1996; see however Perazzona et al. 2004; for general reviews on CREB and memory see Pittenger and Kandel 2003).

Memory suppressor or enhancer genes (Abel et al. 1998) like CREB isoforms and others are relevant genes in the light of the evolutionary biology of learning. Possibly, the differences in learning between parasitic wasps like *C. glomerata* and *C. rubecula* are correlated with differences in expression levels of such genes, i.e., a fast learner like *C. glomerata* could have a relatively low level of memory suppressor-gene expression and a slow learner like *C. rubecula* could have relatively high levels of memory suppressor-gene expression. Measuring of gene expression in these non-model organisms, of which the genome is not sequenced, is time-consuming, but feasible since the homologous sequences from a few insect species are now available. Genes that have been linked to a certain phenotype in one organism can be used as so-called candidate genes to investigate the mechanism and evolution of similar phenotypes in another species (Fitzpatrick et al. 2005).

CONCLUSION

The CREB gene has now been sequenced in *C. glomerata* and *C. rubecula* (H.M. Smid et al. unpublished data) and putative enhancer and suppressor isoforms have been found analogous to the isoforms found in the fruit fly and the honeybee (Yin et al. 1995b; Eisenhardt et al. 2003). Together with the characterization of the neuronal networks involved in associative learning of parasitic wasps, such as the VUM neurons and the olfactory pathways in *Cotesia*, this work may resolve mechanisms and genes that are linked to natural differences in learning. This would allow us to raise and test new hypotheses on the evolution of learning of a range of other wasp species that occur in many different ecological contexts. Moreover, since the mechanisms involved in learning are conserved, at least at the cellular level, these results will be relevant for the understanding of learning in higher animals and man.

ACKNOWLEDGEMENTS

I would like to thank Maartje Bleeker, Joop van Loon and Louise Vet for valuable discussions on parasitoid learning, and the organization committee for their invitation to participate in the Frontis Workshop, *Chemical communication: from gene to ecosystem.*

REFERENCES

Abel, T., Martin, K.C., Bartsch, D., et al. 1998. Memory suppressor genes: inhibitory constraints on the storage of long-term memory. *Science,* 279 (5349), 338-341.

Bernays, E.A., 1996. Selective attention and host-plant specialization. *Entomologia Experimentalis et Applicata,* 80 (1), 125-131.

Bernays, E.A. and Wcislo, W.T., 1994. Sensory capabilities, information processing, and resource specialization. *Quarterly Review of Biology,* 69 (2), 187-204.

Bitterman, M.E., Menzel, R., Fietz, A., et al. 1983. Classical conditioning of proboscis extension in honeybees (*Apis mellifera*). *Journal of Comparative Psychology,* 97 (2), 107-119.

Bleeker, M.A., Smid, H.M., Van Aelst, A.C., et al. 2004. Antennal sensilla of two parasitoid wasps: a comparative scanning electron microscopy study. *Microscopy Research and Technique,* 63 (5), 266-273.

Bleeker, M.A.K., Smid, H.M., Steidle, J.L.M., et al. in press-a. Differences in memory dynamics between two closely related parasitoid wasp species. *Animal Behavior.*

Bleeker, M.A.K., Van der Zee, B. and Smid, H.M., in press-b. Octopamine-like immunoreactivity in the brain and suboesophageal ganglion of two parasitic wasps, *Cotesia glomerata* and *Cotesia rubecula. Animal Biology.*

Collett, T.S. and Collett, M., 2002. Memory use in insect visual navigation. *Nature Reviews Neuroscience,* 3 (7), 542-552.

Corfas, G. and Dudai, Y., 1989. Habituation and dishabituation of a cleaning reflex in normal and mutant *Drosophila. Journal of Neuroscience,* 9 (1), 56-62.

Davies, C.R. and Gilbert, N., 1985. A comparative study of the egg-laying behavior and larval development of *Pieris rapae* and *Pieris brassicae* on the same host plants. *Oecologia,* 67 (2), 278-281.

Dukas, R., 1998a. Constraints on information processing and their effects on behavior. *In:* Dukas, R. ed. *Cognitive ecology: the evolutionary ecology of information processing and decision making.* The University of Chicago Press, Chicago, 89-119.

Dukas, R., 1998b. Evolutionary ecology of learning. *In:* Dukas, R. ed. *Cognitive ecology: the evolutionary ecology of information processing and decision making.* The University of Chicago Press, Chicago, 129-164.

Dukas, R., 1999. Costs of memory: ideas and predictions. *Journal of Theoretical Biology,* 197 (1), 41-50.

Dukas, R. and Kamil, A.C., 2001. Limited attention: the constraint underlying search image. *Behavioral Ecology,* 12 (2), 192-199.

Eisenhardt, D., in press. Learning and memory formation in the honeybee (*Apis mellifera*) and its dependency on the cAMP-protein kinase A pathway. *Animal Biology.*

Eisenhardt, D., Friedrich, A., Stollhoff, N., et al. 2003. The AmCREB gene is an ortholog of the mammalian CREB/CREM family of transcription factors and encodes several splice variants in the honeybee brain. *Insect Molecular Biology,* 12 (4), 373-382.

Fitzpatrick, M.J., Ben-Shahar, Y., Smid, H.M., et al. 2005. Candidate genes for behavioural ecology. *Trends in Ecology and Evolution,* 20 (2), 96-104.

Geervliet, J.B.F., Vet, L.E.M. and Dicke, M., 1994. Volatiles from damaged plants as major cues in long-range host-searching by the specialist parasitoid *Cotesia rubecula. Entomologia Experimentalis et Applicata,* 73 (3), 289-297.

Geervliet, J.B.F., Vet, L.E.M. and Dicke, M., 1996. Innate responses of the parasitoids *Cotesia glomerata* and *C. rubecula* (Hymenoptera: Braconidae) to volatiles from different plant-herbivore complexes. *Journal of Insect Behavior,* 9 (4), 525-538.
Geervliet, J.B.F., Vreugdenhil, A.I., Dicke, M., et al. 1998. Learning to discriminate between infochemicals from different plant-host complexes by the parasitoids *Cotesia glomerata* and *C. rubecula. Entomologia Experimentalis et Applicata,* 86 (3), 241-252.
Giurfa, M., 2003. Cognitive neuroethology: dissecting non-elemental learning in a honeybee brain. *Current Opinion in Neurobiology,* 13 (6), 726-735.
Hammer, M., 1993. An identified neuron mediates the unconditioned stimulus in associative olfactory learning in honeybees. *Nature,* 366 (6450), 59-63.
Hammer, M., 1997. The neural basis of associative reward learning in honeybees. *Trends in Neurosciences,* 20 (6), 245-252.
Hammer, M. and Menzel, R., 1998. Multiple sites of associative odor learning as revealed by local brain microinjections of octopamine in honeybees. *Learning and Memory,* 5 (1/2), 146-156.
Huber, K.M., Kayser, M.S. and Bear, M.F., 2000. Role for rapid dendritic protein synthesis in hippocampal mGluR-dependent long-term depression. *Science,* 288 (5469), 1254-1257.
Krasne, F.B. and Glanzman, D.L., 1995. What we can learn from invertebrate learning. *Annual Review of Psychology,* 46, 585-624.
Kreissl, S., Eichmuller, S., Bicker, G., et al. 1994. Octopamine-like immunoreactivity in the brain and subesophageal ganglion of the honeybee. *Journal of Comparative Neurology,* 348 (4), 583-595.
Le Masurier, A.D., 1994. Costs and benefits of egg clustering in *Pieris brassicae. Journal of Animal Ecology,* 63 (3), 677-685.
Lutz, J., 1994. *Introduction to learning and memory.* Brooks/Cole Pub. Co., Pacific Grove.
Menzel, R., 1999. Memory dynamics in the honeybee. *Journal of Comparative Physiology. A. Sensory Neural and Behavioral Physiology,* 185 (4), 323-340.
Menzel, R., Manz, G., Menzel, R., et al. 2001. Massed and spaced learning in honeybees: the role of CS, US, the intertrial interval, and the test interval. *Learning and Memory,* 8 (4), 198-208.
Mery, F. and Kawecki, T.J., 2003. A fitness cost of learning ability in *Drosophila melanogaster. Proceedings of the Royal Society of London. Series B. Biological Sciences,* 270 (1532), 2465-2469.
Mery, F. and Kawecki, T.J., 2005. A cost of long-term memory in *Drosophila. Science,* 308 (5725), 1148.
Michel-Salzat, A. and Whitfield, J.B., 2004. Preliminary evolutionary relationships within the parasitoid wasp genus *Cotesia* (Hymenoptera: Braconidae: Microgastrinae): combined analysis of four genes. *Systematic Entomology,* 29 (3), 371-382.
Nguyen, P.V., Abel, T. and Kandel, E.R., 1994. Requirement of a critical period of transcription for induction of a late phase of LTP. *Science,* 265 (5175), 1104-1107.
Pavlov, I.P., 1927. *Conditioned reflexes: an investigation of the physiological activity of the cerebral cortex.* Oxford University Press, Oxford.
Perazzona, B., Isabel, G., Preat, T., et al. 2004. The role of cAMP response element-binding protein in *Drosophila* long-term memory. *Journal of Neuroscience,* 24 (40), 8823-8828.
Pittenger, C. and Kandel, E.R., 2003. In search of general mechanisms for long-lasting plasticity: Aplysia and the hippocampus. *Philosophical Transactions of the Royal Society of London. B. Biological Sciences,* 358 (1432), 757-763.
Poolman-Simons, M.T.T., Suverkropp, B.P., Vet, L.E.M., et al. 1992. Comparison of learning in related generalist and specialist eucoilid parasitoids. *Entomologia Experimentalis et Applicata,* 64 (2), 117-124.
Potting, R.P.J., Otten, H. and Vet, L.E.M., 1997. Absence of odour learning in the stemborer parasitoid *Cotesia flavipes. Animal Behaviour,* 53 (6), 1211-1223.
Rescorla, R.A., 1988. Behavioral studies of Pavlovian conditioning. *Annual Review of Neuroscience,* 11, 329-352.
Roitberg, B.D., Reid, M.L. and Li, C., 1993. Choosing hosts and mates: the value of learning. *In:* Papaj, D.R. and Lewis, A.C. eds. *Insect learning: ecological and evolutionary perspectives.* Chapman & Hall, New York, 174-194.
Root, R.B. and Kareiva, P.M., 1984. The search for resources by cabbage butterflies (*Pieris rapae*): ecological consequences and adaptive significance of markovian movements in a patchy environment. *Ecology,* 65 (1), 147-165.

Sandoz, J.C., Laloi, D., Odoux, J.F., et al. 2000. Olfactory information transfer in the honeybee: compared efficiency of classical conditioning and early exposure. *Animal Behaviour,* 59 (5), 1025-1034.

Schröter, U., 2002. *Aufsteigende Neurone des Unterschlundganglions der Biene: Modulatorische neurone und sensorische Bahnen.* PhD Dissertation, Free University of Berlin.

Schwaerzel, M., Monastirioti, M., Scholz, H., et al. 2003. Dopamine and octopamine differentiate between aversive and appetitive olfactory memories in *Drosophila. Journal of Neuroscience,* 23 (33), 10495-10502.

Shettleworth, S.J., 1993. Varieties of learning and memory in animals. *Journal of Experimental Psychology Animal Behavior Processes,* 19 (1), 5-14.

Silva, A.J., Kogan, J.H., Frankland, P.W., et al. 1998. CREB and memory. *Annual Review of Neuroscience,* 21, 127-148.

Skinner, B.F., 1938. *The behavior of organisms: an experimental analysis.* Appleton-Century, New York.

Smid, H.M., Bleeker, M.A., Van Loon, J.J., et al. 2003. Three-dimensional organization of the glomeruli in the antennal lobe of the parasitoid wasps *Cotesia glomerata* and *C. rubecula. Cell and Tissue Research,* 312 (2), 237-248.

Smid, H.M., Van Loon, J.J.A., Posthumus, M.A., et al. 2002. GC-EAG-analysis of volatiles from Brussels sprouts plants damaged by two species of *Pieris caterpillars*: olfactory receptive range of a specialist and a generalist parasitoid wasp species. *Chemoecology,* 12 (4), 169-176.

Stephens, D.W., 1993. Learning and behavioural ecology: incomplete information and environmental predictability. *In:* Papaj, D.R. and Lewis, A.C. eds. *Insect learning: ecological and evolutionary perspectives.* Chapman & Hall, New York, 195-217.

Takasu, K. and Lewis, W.J., 2003. Learning of host searching cues by the larval parasitoid *Microplitis croceipes. Entomologia Experimentalis et Applicata,* 108 (2), 77-86.

Thorndike, E.L., 1901. Animal intelligence: an experimental study of the associative processes in animals. *Psychological Review Monograph Supplement,* 2, 1-109.

Turlings, T.C.J., Wäckers, F.L., Vet, L.E.M., et al. 1993. Learning of host-finding cues by hymenopterous parasitoids. *In:* Papaj, D.R. and Lewis, A.C. eds. *Insect learning: ecological and evolutionary perspectives.* Chapman & Hall, New York, 51-79.

Vet, L.E.M. and Dicke, M., 1992. Ecology of infochemical use by natural enemies in a tritrophic context. *Annual Review of Entomology,* 37 (1), 141-172.

Vet, L.E.M., Lewis, W.J. and Cardé, R.T., 1995. Parasitoid foraging and learning. *In:* Cardé, R.T. and Bell, W.J. eds. *Chemical ecology of insects. vol. 2.* Chapman & Hall, New York, 65-101.

Vet, L.E.M., Lewis, W.J., Papaj, D.R., et al. 1990. A variable-response model for parasitoid foraging behavior. *Journal of Insect Behavior,* 3 (4), 471-490.

Watanabe, H., Kobayashi, Y., Sakura, M., et al. 2003. Classical olfactory conditioning in the cockroach *Periplaneta americana. Zoological Science,* 20 (12), 1447-1454.

Yin, J.C., Del Vecchio, M., Zhou, H., et al. 1995a. CREB as a memory modulator: induced expression of a dCREB2 activator isoform enhances long-term memory in *Drosophila. Cell,* 81 (1), 107-115.

Yin, J.C. and Tully, T., 1996. CREB and the formation of long-term memory. *Current Opinion in Neurobiology,* 6 (2), 264-268.

Yin, J.C., Wallach, J.S., Del Vecchio, M., et al. 1994. Induction of a dominant negative CREB transgene specifically blocks long-term memory in *Drosophila. Cell,* 79 (1), 49-58.

Yin, J.C., Wallach, J.S., Wilder, E.L., et al. 1995b. A *Drosophila* CREB/CREM homolog encodes multiple isoforms, including a cyclic AMP-dependent protein kinase-responsive transcriptional activator and antagonist. *Molecular Cell Biology,* 15 (9), 5123-5130.

CHAPTER 8

VISUALIZING A FLY'S NOSE

Genetic and physiological techniques for studying odour coding in Drosophila

MARIEN DE BRUYNE

Biological Sciences, Monash University, Wellington Road, Clayton VIC 3800, Australia. E-mail: Marien.DeBruijne@sci.monash.edu.au

Abstract. Most insect species rely on odours to orient themselves towards resources or escape hazardous environments. Over the past six years studies on odour perception in *Drosophila melanogaster* have rapidly increased our knowledge on the detection of such signals. Due to the availability of relatively straightforward genetic techniques, the cellular elements of the olfactory code in this insect can be manipulated. Olfactory receptor neurons (ORN) in *Drosophila* can be visualized with fluorescent proteins and their physiological properties studied using electrophysiological and optophysiological techniques. The ultrastructure of olfactory sensilla and the odour responses of ORNs in more than half of them have been described. On the molecular level, three large families of genes that provide the basis for these responses have been characterized; olfactory receptors (OR), gustatory receptors (GR) and odour-binding proteins (OBP). OR proteins have been shown to function as odour detectors and they have been mapped to ORN classes to a high degree of completion. Hence, the *Drosophila* olfactory system provides a good basis for studying how odour coding in insects has evolved and how ORNs relay the information present in chemical communication systems.
Keywords: olfaction; *Drosophila*; genetics; sensory physiology; neural coding

INTRODUCTION

Chemical signals are involved in most interactions of insects with their environment. Volatile chemicals (i.e., odours) are signals that have many degrees of freedom and can travel far. Some, such as sex pheromones, can be specific, stable predictors of reproductive success. Because both signal and response are generated by the same genome, highly specialized systems for pheromone synthesis and perception have evolved (Löfstedt 1993). However, most odours are not generated by conspecifics but rather by a large variety of biotic and abiotic factors. In fact, many odours are the result of complex interactions, as for example in weather-dependent microbial decay of plant material. How have these sensory systems evolved to extract reliable chemical information from variable environments? Have olfactory systems evolved as a set of detectors for specific chemical messages or are they designed for efficient detection of a broad range of chemical stimuli? To answer these questions we need

M. Dicke and W. Takken (eds.), Chemical Ecology: From Gene to Ecosystem, 105-125.

to determine how a complete olfactory system works, first in one species and then in a comparative way across species.

Encoding of odour information is a two-step process. First, sensory transduction converts chemical information in the environment into a code of action potentials. It takes place in a heterogeneous population of olfactory receptor neurons (ORNs) that determine which volatiles can be detected. Second, from the messages sent out by this array of detectors the brain extracts a percept we call an odour. It is this combined input from many ORN classes that can lead to a behavioural response, depending on the animal's internal state and the integration with other sensory modalities. Most recent research has focused on the first step of this process. The process of capturing and transducing chemical information from the environment was thought to involve G-protein-coupled receptors (Boekhoff et al. 1990), but convincing evidence was lacking. Buck and Axel (Buck and Axel 1991) made a crucial breakthrough when they discovered a large gene family encoding such receptors in vertebrates. Evidence for their crucial role in transducing olfactory information came from studies in *C. elegans* (Sengupta et al. 1996). It was only after genomic sequences of *Drosophila melanogaster* became available that candidate odour receptor proteins were identified in an insect (Clyne et al. 1999; Vosshall et al. 1999). This paper will argue that research on *Drosophila* olfaction has significantly advanced our knowledge on the mechanisms of olfactory perception and should also help in answering more ultimate questions about the ecology and evolution of chemical communication. I will provide an overview of the powerful techniques available in this model organism.

DROSOPHILA OLFACTORY ORGANS

Drosophila melanogaster has rapidly become the favourite model system for studying olfactory coding (Carlson 1996; Vosshall 2000; Stocker 2001; Davis 2004). The reasons for this are many. Its olfactory system is numerically simple, containing only ca. 1300 receptor neurons (Stocker 1994). Furthermore, there are powerful genetic and molecular tools to manipulate the system and determine its genetic underpinnings. Moreover, the *Drosophila* genome has been sequenced and annotated very accurately. Several physiological and genetic techniques are available to peer into the workings of the little fly's sensory organs and associated neuropiles in the central nervous system. Great progress has been made in visualizing neuronal structures and studying neural activity (De Bruyne et al. 1999; 2001; Jefferis et al. 2002; Fiala et al. 2002; Ng et al. 2002; Wang et al. 2003; Wilson et al. 2004). Finally and perhaps most importantly, physiological and genetic analysis can be combined with simple assays for innate or conditioned behaviour.

Drosophila has a relatively simple olfactory system with ORNs distributed over two paired appendages, the antennae, which carry most of the receptor neurons, and maxillary palps (Figure 1A, Stocker 1994; De Bruyne 2003). The *Drosophila* antenna does not have a segmented flagellum like most other insects. Instead all olfactory sensilla are on one segment that does not contain taste or mechanosensory

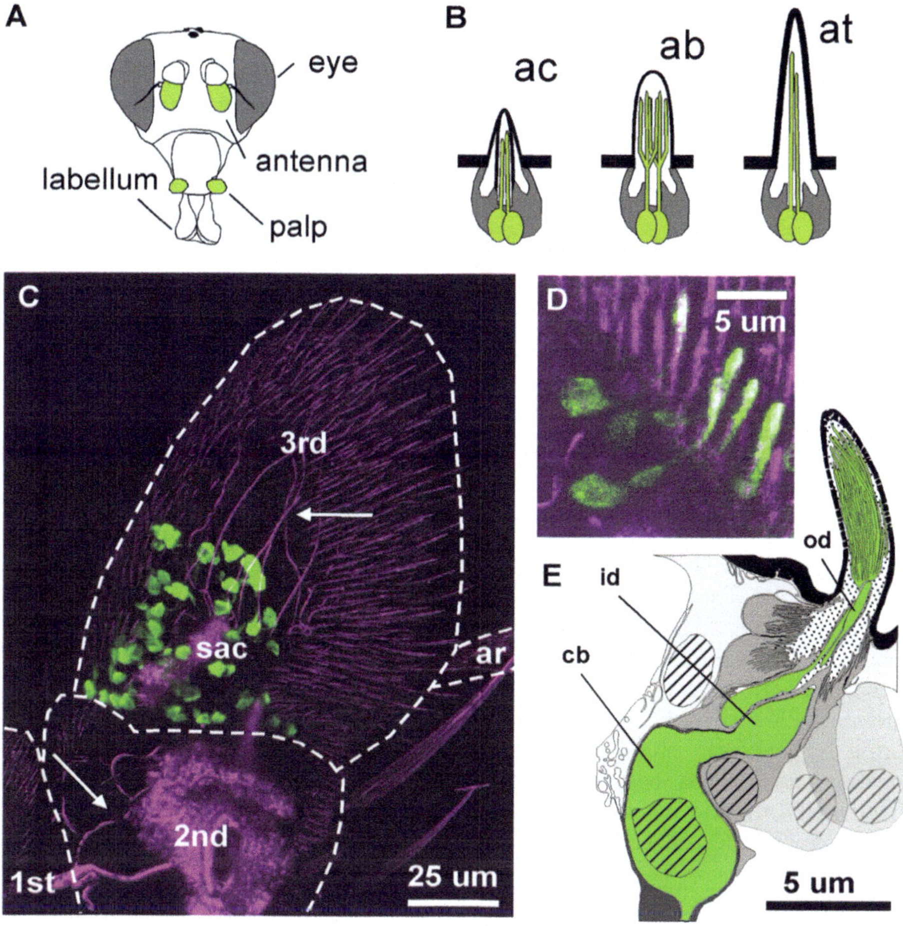

Figure 1. Visualizing olfactory receptor neurons (ORN) of Drosophila. A. Drosophila head with main sensory organs. ORNs (green) can be found on the antennal third segment (funiculus) and the maxillary palp. B. ORNs on the antenna are housed in sensilla made up of a cuticular hair or peg with a pored wall, 1-4 neurons (green) and 3-4 accessory cells (grey, see also under E). There are three structural categories of sensilla: antennal coeloconics (ac), basiconics (ab) and trichoids (at). C. A confocal image of an antenna with ab3A neurons labelled by membrane-bound mCD8::GFP (green) dr iven by Or22a-Gal4, the regulatory region of a receptor gene. Medial view of three antennal segments (1st, 2nd and 3rd) with cuticular structures visualized by reflected light (magenta). Sac, sacculus; ar; arista. Arrows point to trachaea. D. Detail of GFP-labelled receptor neurons innervating basiconic sensilla. E. Cellular components of a typical basiconic sensillum. Neurons in green, accessory cells in grey, glial cell in dark grey. Epidermal cells are light grey. Note the thin outer dendrite with branches filling the sensillum shaft (od), spindle-shaped inner dendrite (id) and round cell body (cb) very similar to the neurons in D. Drawn to scale after Shanbhag et al. (2001)

sensilla. In most insects, gustatory receptor neurons (GRN) are mixed with ORNs on the antennae, but *Drosophila* offers the advantage that GRNs are found only on other appendages such as mouthparts and legs (Stocker 1994). As in all insects, the ORNs are housed in sensilla made up of small sets of epithelial cells (Figure 1B). A sensillum is composed of three elements (Figure 1E). First there is a cuticular apparatus, usually a hair or short peg with a pored wall. The accessory cells make up the second element. They supply the hair with a lymph that surrounds the dendrites of the last element; the neurons themselves. *Drosophila* olfactory sensilla contain 1-4 neurons that send their dendrites into the hair and their axons to the antennal lobe in the brain. ORNs of a single class converge on a single member out of a set of ca. 40 glomeruli (Stocker 1994; Laissue et al. 1999; Vosshall 2000, see also Figure 2A); small spherical sub-regions with a high density of synaptic contacts between ORNs, local interneurons and projection neurons. Both the palp and the antennal third segment are small (<100μm) and nearly transparent organs so their sensilla can be visualized under high magnification in a compound microscope. The antennal sensilla fall into two ultrastructural categories, double-walled (dw) and single-walled (sw) (Altner and Prillinger 1980). The dw sensilla of *Drosophila* are known as 'coeloconic' sensilla (Figure 1B, Venkatesh et al. 1984) and have a different developmental origin (Gupta and Rodrigues 1997; De Bruyne 2003) compared to the sw sensilla. They are not – as the term 'coeloconic' implies – situated in pits but merely in slight depressions which can be clearly seen under the microscope as circles. The sw sensilla are more abundant and have traditionally been further categorized as 'basiconic' and 'trichoid' sensilla (Figure 1B, Venkatesh et al. 1984): short peg-shaped with a rounded tip or longer and more hair-like, respectively. Around 550 of all 1200 neurons on the antenna are in basiconic sensilla. The maxillary palp bears only 60 basiconic sensilla housing 120 neurons in pairs. Compared to some other insect species that are important models in olfaction (moths, bees, cockroaches, locusts) *Drosophila* has fewer neurons that are more easily visualized.

CHEMOSENSORY GENES

Three large families of genes providing the molecular basis for detection of chemicals have been characterized: olfactory receptors (OR), gustatory receptors (GR) and odour-binding proteins (OBP). The first members of the *Drosophila* OR gene family were isolated by searching through genomic DNA sequences using algorithms designed to pull out sequences coding for multiple transmembrane domains in the predicted protein (Clyne et al. 1999; Vosshall et al. 1999). The sequencing of the full *Drosophila* genome sequence has revealed 60 OR genes with highly divergent sequences (Vosshall et al. 2000; Robertson et al. 2003). They show no sequence similarity to those of vertebrates or nematodes. However, a common feature of ORs across phyla is that they belong to the superfamily of seven transmembrane domain G-protein-coupled receptors (GPCRs). A second family of

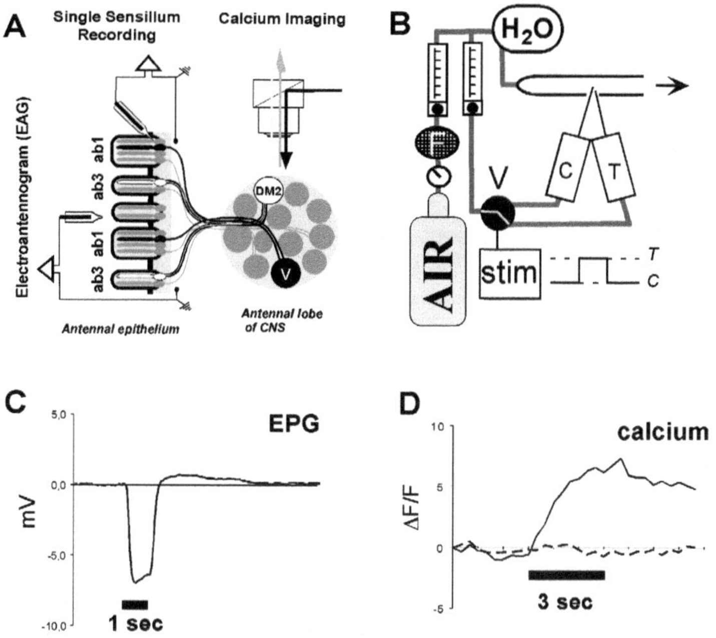

Figure 2. *Recording from olfactory receptor neurons.* ***A.*** *Schematic view of three ways to record activity in olfactory receptor neurons (ORNs). Two classes of Drosophila ORNs are indicated here: ab1C neurons (black), which send their axons to the V glomerulus, and ab3A neurons (white), which project to the DM2 glomerulus (seetext). In single-sensillum recordings (SSR) electrical activity is measured by bringing an electrode (glass or tungsten) into contact with the lymph of a single sensillum. Electroantennograms (EAG) or eletro-palpograms (EPG) measure changes in the transepithelial potential by depositing a glass electrode on the cuticle. It presumably measures the combined activity of many sensilla but we do not know exactly how this process is accomplished. Finally, neural activity can be assessed by optical measurements on fluorescent-calcium sensors genetically targeted to certain receptor neurons. This can be done through the cuticle of the antenna or (as shown here) on the exposed antennal lobe where all ORNs of a single class converge onto glomeruli (see text) (Fiala et al. 2002).* ***B.*** *Odour stimulation is by delivering controlled pulses of odour-laden air into a constant air stream over the preparation. Air is charcoal-filtered (F), humidified (H_2O) and blown at relatively high speed (180 cm/s) from a glass tube with a small hole in the sidewall. Two syringes have their needles inserted through this hole. One of them (C) is empty and adds a constant flow of clean air. Odours are dis-solved in paraffin oil on filter paper placed in another syringe (T). A valve (V), electrically regulated by a stimulator (stim), switches the flow briefly from C to T adding odour to the air without disturbing other properties such as speed, turbulence, humidity or temperature.* ***C.*** *Example of an EPG response to ethyl acetate.* ***D.*** *Example of calcium responses recorded from ab5B neurons expressing cameleon. ΔF/F indicates the increase in the ratio of fluore-scence from its two fluorophores relative to the background fluorescence. Pentyl acetate (black line) evokes a change in calcium concentration not observed in a paraffin-oil control (dashed line). Baseline instabilities were correted for by subracting responses to clean air*

such genes was described soon after (Clyne et al. 2000; Dunipace et al. 2001; Scott et al. 2001). These were named gustatory receptors (GR) because most of them are expressed in taste sensilla. The sequences of this family are even more diverse. In *Drosophila* some GR genes are expressed in antennae and may well have an olfactory function as there are no taste sensillae there. The OR genes are thought to have evolved from a subfamily of GR genes (Robertson et al. 2003). Individual members of the OR family are expressed in small subsets of ORNs, with different members expressed in different subsets (Clyne et al. 1999; Vosshall et al. 2000). As in vertebrates, axons of ORNs expressing a particular OR gene converge onto single glomeruli (Vosshall et al. 2000; Gao et al. 2000).

Members of a third gene family that is thought to play a role in mediating odour response variability are generally referred to as odour-binding proteins (OBP). They are not membrane-bound and not neuronal but secreted in large quantities into the extracellular lymph by accessory cells of sensilla or by epithelial cells (Shanbhag et al. 2001). Unlike Ors, the OBPs are not produced exclusively in olfactory sensilla. Nevertheless, they represent a varied set of genes that are differentially expressed in olfactory tissues. This has long been taken as strongly indicative for a role in determining response properties of ORN (Steinbrecht et al. 1995). Evidence for this has recently come from a mutation in one of the *Drosophila* OBPs (obp76a) called *lush* (Xu et al. 2005).

USING THE *DROSOPHILA* TOOLKIT TO STUDY OLFACTORY CODING

Drosophila genetics

The main reason to use *Drosophila* as a model species is of course its amenability to genetic manipulation. There are two ways to study a biological system in *Drosophila* genetically. Classical or 'forward' genetics starts from randomly induced mutations, causing changes in the fly's phenotype that are of interest to the questions at hand. By exposing flies to certain chemicals or radiation, changes in the DNA sequence can be induced (mutagenesis). These mutations are then 'mapped' to a locus in the genome, ideally to a single gene. Nowadays geneticists more often work the other way round. They have identified a candidate gene from the *Drosophila* genome and want to know its effect on a certain phenotype. Attempts to target a particular gene by either reducing or enhancing its function are referred to as reverse genetics.

Several reverse-genetic techniques use transposons, small pieces of DNA that occur naturally and have the property of moving around in the genome by removing and reinserting themselves. The so-called P element is most commonly used. The coding region for the transposase, an enzyme that mediates its mobility, has been removed and a genetic marker gene has been added so that its presence in the genome can be seen in a fly's phenotype. These transposons can be genetically engineered and used to incorporate foreign DNA (transgenes) in the *Drosophila* genome.

The Gal4/UAS system makes it possible to express a transgene in a specific tissue only. Gal4 is a gene originally found in yeast that has no normal function in

Drosophila cells. It encodes a transcription factor; a protein that regulates other yeast genes by binding to DNA at a site called UAS (upstream activating sequence). Brand and Perrimon (1993) cloned the Gal4 gene and the UAS sequence in two separate P-elements, introduced in two separate parental fly lines. One fly line contains the 'driver' element, expressing Gal4 in a particular set of cells and/or at a particular time in development. The other line contains the 'responder' element; the gene of choice under the control of the UAS. When the two are crossed, Gal4 can bind to UAS and activate expression of its transgene, but only in the targeted cells that express Gal4. There are many different Gal4 lines with highly specific expression patterns. Similarly, many useful genes have been introduced in UAS elements. This highly flexible expression system is now widely used as a tool to target specific cell populations such as ORN classes or accessory cells in flies.

In a certain GAL4/UAS approach a P element incorporating Gal4 with a weak promoter is allowed to 'jump' around in the genome by crossing in a chromosome carrying another P element with the transposase. It is then thought to insert randomly, and the transgene is expressed under the influence of regulatory sequences close to the insertion point. Such 'enhancer trap lines' have several disadvantages. For instance, out of thousands of random insertions that were scanned for expression in the brain, very few expression patterns are specific for a particular class of cells (Ito et al. 2003). A more reliable approach is to generate specific Gal4 constructs that include regulatory sequences (promoter and enhancer elements) upstream of a known gene. Such 'promoter constructs' ideally place the Gal4 under the control of the regulatory region of the chosen gene to target the UAS transgene specifically to cells that express it. However, even in this case, care must be taken. The expression pattern can differ from the actual gene's expression because not all enhancing elements were included or suppressing elements were omitted. In addition, Gal4 expression can depend on the insertion point of the P element.

Genetic tools for manipulating olfactory neurons

To target ORNs the most specific sequences that can be used to drive Gal4 expression are those regulating expression of OR genes. A number of OR promotor–Gal4 constructs have been made, which drive expression of Gal4 in a small subset of ORNs (Vosshall et al. 2000; Goldman et al. 2005; Kreher et al. 2005). Several readily available UAS constructs allow the expression of transgenes that visualize cells, allow neuronal-activity monitoring or the inactivation of cells, either permanently or under certain conditions. The gene for green-fluorescent protein (GFP), originally from a jellyfish, has been manipulated to render a cytosolic protein that does not damage the cells and gives strong green (510 nm) fluorescence upon excitation with blue (488 nm) light. It has been coupled to the mouse gene CD8 to give a fusion protein that localizes to the cell membrane (Lee and Luo 1999). Fluorescence can be observed in living flies under a stereomicroscope or in whole mounts under a confocal laser scanning microscope (Figure 1C,D) and the full shape of ORNs can be resolved.

Apart from making cells visible, proteins can be expressed that indicate cellular function. Several UAS transgenes make proteins that change their fluorescent properties with neural activity. Two of these measure intracellular calcium concentrations that usually go up when neurons are depolarized: Cameleon (Fiala et al. 2002) and GcamP (Wang et al. 2003). A third construct, synapto-pHluorin, is localized in synaptic vesicles and increases its fluorescence when neurotransmitters are released during synaptic transmission (Ng et al. 2002).

All these proteins are presumed to have very little influence on the physiology of the cell. Cell function itself can also be manipulated in various ways. A modified version of the bacterial gene for tetanus toxin is used to block synaptic transmission, effectively eliminating neuronal function (Sweeney et al. 1995). Expression of the *rpr* and/or *hid* genes can selectively ablate cells since these genes are part of the cellular mechanism for programmed cell death (Bergmann et al. 1998). One disadvantage of killing or disabling cells is that this can disturb development during embryogenesis or metamorphosis. An alternative is the use of constructs that are only activated under certain conditions. Such as the temperature-sensitive *UAS–Shi* construct (Kitamoto 2001). *Shi* (shibire) is a mutation in the dynamin protein, which is involved in synaptic-vesicle recycling. Expression of this protein is harmless at 25°C, but at 33°C the mutated form effectively blocks the native form and synaptic transmission stops. The advantage is that the flies can develop and behave normally until the temperature is raised during a particular experiment.

Recording neuronal activity

Drosophila's small size can be a challenge for physiologists but also offers distinct advantages. The antennal epithelium is packed with neurons and has a thin, transparent cuticle, so one can see a large set of olfactory sensilla in a single view under a compound microscope. There are three techniques available for recording neuronal responses to olfactory stimuli (Figure 2A). Detailed information on ORN responses can be obtained from single sensillum recordings (SSR), but these are laborious and technically demanding to perform. A quick but less informative approach is the electroantennogram (EAG), where an integrated response from many ORNs is recorded. Finally, specific GAL4 constructs that label ORNs can be used to drive calcium-sensitive proteins in order to measure neuronal activity optically. A reliable and flexible odour delivery system is used, which minimizes contamination between stimuli (Figure 2B). A glass tube continuously supplies humidified air to the preparation while two syringes are inserted through a small hole. A second flow of air (or nitrogen) passes constantly through an empty syringe and can be temporarily switched by a computer-controlled solenoid valve to push odorants from an odour-laden syringe. The delay time in the physiological response after activating the switch is determined partly by the airspeed and the distance from the injection point to the preparation and partly by the physiological response latency. The latter is around 20 ms and can be attributed to physico-chemical events at the air–liquid and liquid–membrane interfaces, as well as the transduction cascade inside the neurons.

Electrophysiological recordings are made from single olfactory sensilla by bringing an electrode (a saline-filled glass capillary with a silver wire or an electrolytically sharpened tungsten wire) in contact with the liquid surrounding the dendrites (Figure 2A). The reference electrode is inserted in the eye or in the thin cuticular folds at the base of the proboscis. This set-up measures extracellular voltage differences across the epithelium between the haemolymph and the sensillum lymph. Because the electrical resistance between individual sensilla is relatively high, only events that take place in the contacted sensillum are recorded. The relative amplitudes of action potentials fired by different ORNs in a single insect sensillum indicate the number of neurons present, and this phenomenon is used to analyse their activity separately (Kaissling 1995). The spikes of the ORNs in *Drosophila* basiconic sensilla can usually be separated reliably this way (Figure 3A,B, De Bruyne et al. 1999; 2001). When recording with glass electrodes one can also record the so-called sensillum potential (SP), an extra-cellular derivative of the membrane potentials of the neurons as well as the ion-pumping activity of the accessory cells that determines the potential difference between the sensillum lymph and the haemolymph (Figure 3C). Changes in this trans-epithelial potential in response to odorant stimulation are thought to reflect receptor potentials of the neurons (Kaissling and Colbow 1987).

Electroantennograms (EAG) can give reliable and more easily obtained information about which odorants are detected by insect ORNs (Figure 2A,C). Changes in the trans-epithelial potential can be recorded from the whole antenna using various electrode arrangements. From large insects such as moths, EAGs are recorded between the base and the cut tip of a severed antenna (Kaissling 1995). In Diptera the reference electrode is generally inserted in the haemolymph of the head and the recording electrode brought into contact with the antennal surface (Guerin and Visser 1980). To record EAG signals from *Drosophila* the fly is left intact, positioned inside a plastic pipette tip and the recording electrode placed on the medio-proximal part of the third antennal segment (Ayer and Carlson 1992). An equivalent signal can also be recorded from the maxillary palp, the electropalpogram (EPG, Ayer and Carlson 1992). The voltage deflections observed in response to odours are a summation of the SPs of many ORNs. They show very similar dynamics but smaller amplitudes. Although EAG recordings supply excellent resolution when comparing wild type to mutant phenotypes, they do not allow conclusions on odour coding per se, because the exact working principles of the EAG are poorly understood. For instance, it is not known how many ORNs are measured or what their relative contribution to the EAG is.

Optical imaging has been used extensively in recent years to measure activity from insect brains (Galizia and Menzel 2001, and references therein). In *Drosophila* it is now possible to express various sensors that register either intracellular calcium (Fiala et al. 2002; Wang et al. 2003) or changes in the pH of the synaptic cleft (Ng et al. 2002) because they are available as a UAS construct. To visualize signals in the *Drosophila* brain it is invariably necessary to remove the cuticle and expose the brain (Galizia and Vetter 2005). However, because the antennal cuticle is transparent

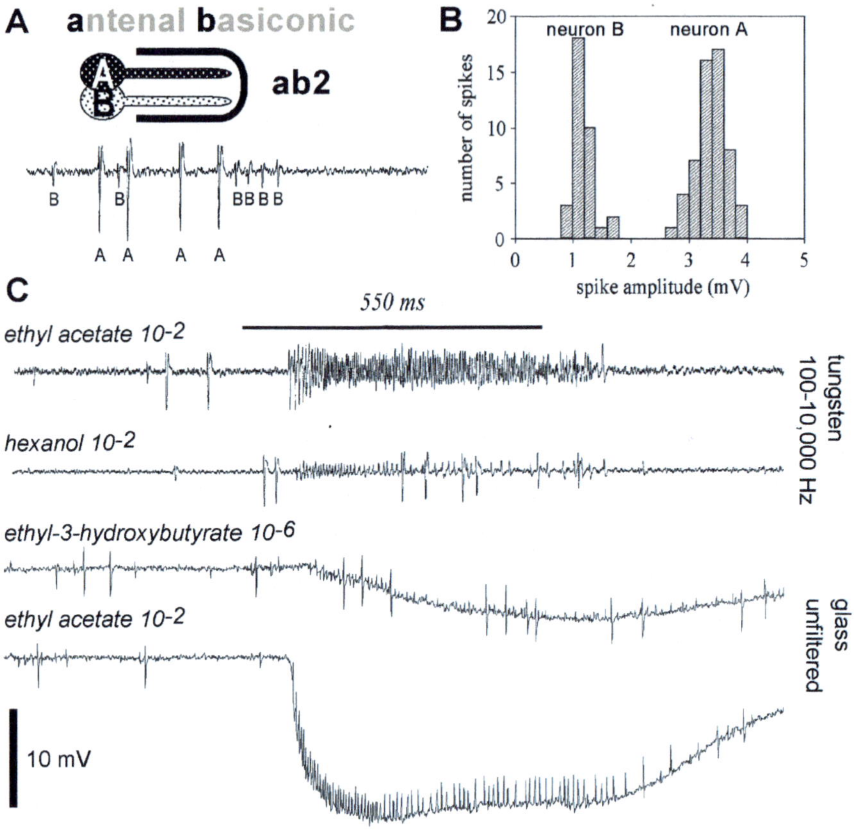

Figure 3. *Single sensillum recording of odour responses in receptor neurons.* ***A.*** *Antennal basiconic sensilla of type ab2 contain two neurons that are identified by size and shape of the action potentials they fire as A (the larger) or B (the smaller).* ***B.*** *Frequency histogram of the spike amplitudes shows a bimodal distribution.* ***C.*** *Traces of recordings from ab2 sensilla using either tungsten electrodes with filtering (top two traces) or glass electrodes to reveal slow sensillum potentials. Odorants were dissolved into 20 µl of paraffin oil on filter paper according to the dilution factors indicated. Note that the B neuron is excited by ethyl-3-hydroxybutyrate at a 10,000x lower concentration than hexanol. Partly after de Bruyne et al. 2001, with permission*

it is possible to image ORNs in an intact fly (Figure 2A,D). One of the proteins engineered to indicate calcium concentration is called cameleon (Fiala et al. 2002). It combines calmodulin (a calcium-binding protein) and the calmodulin-binding domain from myosin (M13) with two modified GFP proteins that emit fluorescent light at either 485 nm (cyan) or 535 nm (yellow). An increase in intracellular calcium, such as occurring during an odour response, results in calmodulin binding to M13 and fluorescent activity moving from the cyan to the yellow GFP. The latter event can be measured by recording light emission at two wavelengths and calculating the ratio. Optophysiological measurements of antennae offer the advantage of being technically less challenging than single sensillum recording. However, the exact relation between calcium signals and action-potential firing rates has not been established and it is of course the latter that drive behavioural output. For instance, calcium signals tend to rise and descend much slower than spike frequencies.

PHYSIOLOGICAL STUDIES REVEAL AN ARRAY OF RESPONSE UNITS

Cell types and distribution

All neural computing that leads to a useful representation of odours in the insect brain must be based on information supplied by receptor neurons. To understand an insect's response to odours, a complete picture of the neural code entering the brain would be desirable. What do typical ORN responses look like and how is neuronal activity in response to a single odorant distributed across all neurons? Figure 3 shows that short stimulations with odours result in increased firing of action potentials. The firing frequency generally increases rapidly (within 100 ms) to a maximum and then falls back due to adaptation (De Bruyne et al. 1999). Most neurons respond to several odorants but with differing sensitivity. Moreover, individual ORNs can be classified into classes with very similar response properties. Recording responses to several odorants from a large number of *Drosophila* basiconic sensilla has revealed 22 different ORN classes (De Bruyne et al. 1999; 2001). Elmore et al. (2003) added 2 more. These 24 neuron classes represent more than 50% of the entire olfactory input to the brain. In trichoid sensilla there are at least a further 6 ORN classes in three different sensillum types (Clyne et al. 1997; Xu et al. 2005). In addition, coeloconic sensilla also house ORNs with response profiles that fall into distinct classes, some responding to small aliphatic acids (Clyne et al. 1997; Park et al. 2002). Thus the *Drosophila* nose is organized in classes of ORNs with distinct response properties, as has also been observed in other insects (e.g., Kaib 1974). The total number of such coding units will probably be around 40-50 because in *Drosophila* ca. 50 glomeruli (including sub-compartments) have been identified (Laissue et al. 1999), 40 OR and a few GR genes are expressed in palps and antennae (Vosshall et al. 2000, and Figure 6) and there is near one-to-one correspondence between ORN classes, OR expression and projection to glomeruli (Vosshall et al. 2000; Goldman et al. 2005; Kreher et al. 2005).

Not only is there consistency in the response properties of ORNs but also in the way they are combined into sensillum types and (at least on the antenna) how sensillum types are distributed over the surface of the antenna. Genetic studies on the development of sensillum morphologies and ORN identities strongly suggest a hierarchy of events determining the layout of the array of odour detectors (for a review see De Bruyne 2003). In spite of this conspicuous pairing there is no

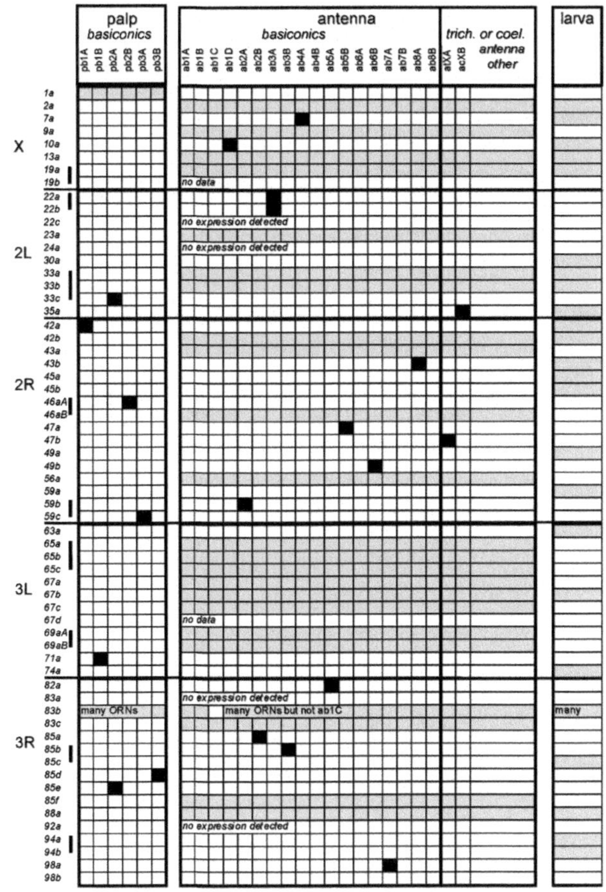

Figure 6. Mapping OR genes to olfactory receptor neuron classes. A summary of recent advances in the mapping of OR gene expression to defined response classes of olfactory receptor neurons or to palp, antenna or whole larva. All 60 genes of the Drosophila OR gene family are listed. The Or46a and Or69a genes are alternatively spliced rendering 62 OR proteins (Robertson et al. 2003). The nomenclature (Warr et al. 2000) indicates a gene's location on either the sex chromosome (X) or one of the arms of the autosomes (2L is left arm of 2nd chromosome). Vertical bars label genes that form small clusters. Identified O RN classes on the maxillary palp (De Bruyne et al. 1999) and antenna (De Bruyne et al. 2001; Elmore et al. 2003) are listed on the top and black squares in the matrix indicate positive mapping of a gene to an identified ORN class in the adult (Hallem et al. 2004; Goldman et al. 2005) and/or to a single larval ORN. Note that Or83b is expressed in most ORNs although its expression has not been verified for each class (Larsson et al. 2004). One identified exception is the ab1C neuron, which does not express Or83b. Grey squares indicate that the gene has been mapped to antenna, palp and/or larva but the ORN class has not been identified (Vosshall et al. 1999; Hallem et al. 2004; Kreher et al. 2005). For some genes data are lacking and for others RT-PCR experiments failed to reveal expression in the two adult olfactory organs (Vosshall et al. 1999) or in larvae (Kreher et al. 2005)

evidence for integration of odour information at the level of sensory neurons. When challenging one of the two neurons of a palpal basiconic sensillum (pb1A) with a strong prolonged stimulation it was found that adaptation of this neuron does not affect the response of the other neuron (De Bruyne et al. 1999). It has been suggested that two ORNs sampling the same μl of air via the same sensillum lymph allows more accurate computation of the concentration ratios of components in a mixture (Todd and Baker 1999). For instance, pairing two neurons, sensitive to the components of a pheromone blend, may allow moths to assess their relative concentrations with a very high spatial and temporal resolution. It is conceivable that this is required for split-second decisions about odour quality during upwind flight. Pheromone plumes are mixed with non-relevant odours from the background, but individual odour packets within such a plume are thought to derive from the same source. If this is a general reason for combining certain ORNs in a single sensillum then we should ask ourselves what the functional relations are between the odorants that cohabiting ORNs detect.

Coding properties

Odour stimuli contain three elements of information that are encoded by the ensemble of ORNs. The first is odour identity, i.e., the chemical structure of a single compound or of several components in a blend. The second is the concentration of the odorants. The third is odour variations in time. Typical experimental odour pulses have an onset and an end. By contrast, natural stimuli consist of variations in odour intensity and identity over time. These variations can be long-term (minutes, hours, days) when insects move from one environment to another, or short-term (milliseconds) as, for instance, when a flying insect traverses an odour plume.

The discussion about identity coding in insect ORNs has focused on whether there are so-called 'specialist' and 'generalist' neurons (Schneider 1984; Hildebrand and Shepherd 1997). In *Drosophila* basiconic sensilla we find examples of both (De Bruyne et al. 1999; 2001). Figure 4 shows how an odour stimulus leads to neural activity across an array of ORN classes. The ab2A neuron would be considered a specialist with its specific response to ethyl acetate. However, classification of odour response spectra as narrowly tuned (specialist) or broadly tuned (generalist) to odours depends very much on the set of stimuli used. The same neuron also responds to acetone and 2,3-butanedione. Moreover, its response to ethyl acetate is not very strong and it is likely that as yet unidentified chemicals provide a better stimulus. The data in Figure 4 show that at these relatively high doses some odours stimulate several ORN classes and some ORN classes respond to more than one odour. The general notion is that odour perception would require the CNS to integrate information across ORNs (combinatorial coding). The ab1C neuron could supply what has been described as a 'labelled line'. In *Drosophila* it is the only ORN that responds to CO_2, and CO_2 is the only odorant it responds to. Other ORNs are also narrowly tuned, e.g., ab1D to methyl salicylate, and ab5A to geranyl acetate (not shown). In a combinatorial code (a.k.a. across-fibre pattern) an odour will be

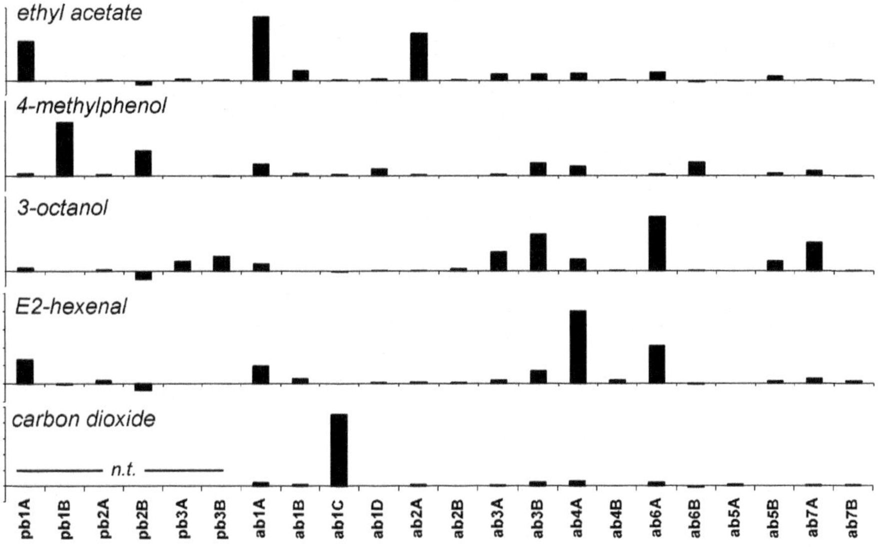

Figure 4. *Odours excite different combinations of receptor neurons. Excitation patterns across 22 olfactory receptor neuron (ORN) classes for five odorants as extrapolated from single-sensillum recordings using 10^{-2} dilutions. Data are from De Bruyne et al. (1999; 2001). pb.... – palpal ORN; ab..... – antennal ORN. Note that CO_2 was not tested (n.t.) on palpal neurons but electropalpograms show no responses*

defined by the combined activity of several ORN classes. In its most extreme form all ORNs would have broad overlapping response spectra and the CNS would be a homogeneous network that converts their activity patterns in odour percepts. The other extreme would be a labelled line system where each odour has a 'dedicated' ORN class defining its perception.

In order to understand principles of odour coding it is important to realize that dose–response relationships are not linear. The typical dose–response curve is sigmoid in shape, rising slowly at lower doses, more or less linear over a range of 2-3 log steps and saturating at spike rates of over 200-300 spikes/s (Figure 5A). As a result, coding of odour identity over a wider range of concentrations would require two or more ORN classes with different sensitivities (De Bruyne 2003). Another consequence of the non-linearity is that odour identity cannot simply be determined from the ratio in firing rates of a set of ORNs because these will be dose-dependent. For our initial characterization of *Drosophila* ORNs, we tested odorants at a relatively high, but not unnatural dose (De Bruyne et al. 2001). When comparing dose–response relations we found that thresholds for the more active stimuli were at least 1000x lower. In order to identify neuron classes one has to find at least one stimulus that elicits a response, but the best ligands for most of these ORNs may not have been in our set of odorants. Stensmyr et al. (2003b) have since identified better stimuli for some of them, such as ethyl-3-hydroxybutyrate, which stimulates the ab2B neuron (Figure 3C).

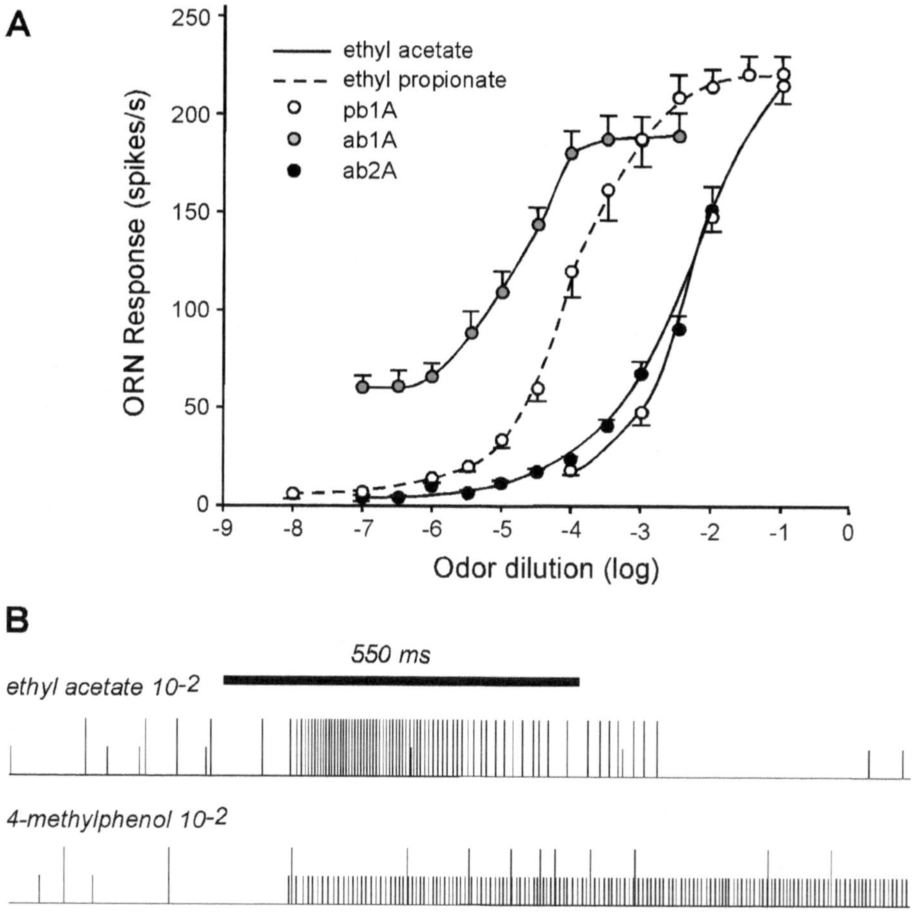

Figure 5. Coding of odour concentration and odour dynamics. ***A.*** *Dose-response curves of olfactory receptor neuron excitation are sigmoid with sensitivities reaching as low as* 10^{-7} *Odour concentration is indicated here as dilutions in 20µl paraffin oil on filter papers placed in odour cartridges. Concentrations reaching the fly are unknown, as they depend on vapour pressure, but air expelled from these cartridges is diluted another 16-fold. Note that at higher doses three neurons respond to ethyl acetate but only one is sensitive to low doses. Meanwhile the pb1A neuron, though excited by ethyl acetate at high doses, responds better to ethyl propionate at low doses.* ***B.*** *Raster plots of neural activity in pb1 sensilla in response to two odorants. Note the strongly phasic-tonic nature of the pb1A neuron's response to ethyl acetate (large spikes in upper trace) and the more tonic and longer-lasting activation of pb1B by 4-methylphenol (small spikes in lower trace). After De Bruyne et al. (1999; 2001), with permission*

The basis of the olfactory code is that ORNs respond with different sensitivities to different odours. However, response kinetics also vary (Figure 5B). The onset of a typical odour pulse induces a sharp rise in firing frequency, which quickly decreases (adaptation), but the end of stimulation is either marked by an abrupt decrease in firing or by a much slower decay in firing. These variations in temporal integration of stimulation are specific for stimulation of a particular ORN with a particular odorant (De Bruyne et al. 2001). Theoretically at least, these properties could contribute to odour coding.

OR GENES: FUNCTIONAL CHARACTERIZATION AND MAPPING TO ORN CLASSES

The predicted role of the *Drosophila* OR genes as odorant receptors has been confirmed by inducing odour responses after the expression of an OR gene and by their removal in case of mutation. One study uses a Gal4 enhancer trap line to over-express the OR43a gene in *Drosophila* ORNs (Störtkuhl and Kettler 2001) while another uses heterologous expression of the same gene in *Xenopus* oocytes (Wetzel et al. 2001). In both cases it was shown that expression of the OR gene leads to physiological responses to specific odorants. Electrophysiological analysis of mutations in the genes Or22a and Or43b showed that odour responses from ORNs that normally express these genes were no longer observed (Dobritsa et al. 2003; Elmore et al. 2003). Each ORN class is restricted to a particular spatial domain on the antennal surface, although there is considerable overlap between them (De Bruyne et al. 2001). The expression patterns of OR genes in the antenna reflect this organization (Clyne et al. 1999; Vosshall et al. 1999; Gao and Chess 1999). Now, the expression of many OR genes has been mapped to the ORN classes (Figure 6) using two different techniques. In the first technique, OR-Gal4 constructs were used to label sensilla with GFP or delete the targeted ORN with *rpr* (Dobritsa et al. 2003; Goldman et al. 2005). Single sensillum recordings were then used to identify the sensillum type. The second technique makes use of the fact that in Or22a mutants the ab3A neuron no longer responds to odorants but it still functions as a neuron (Dobritsa et al. 2003). Other OR genes can be expressed in this 'empty neuron' (Δab3A), inducing odour responses specific for the expressed OR (Hallem et al. 2004; Kreher et al. 2005). The acquired response is then compared to the established spectra of native ORNs.

Expression of OR genes has also been analysed in *Drosophila* larvae (Kreher et al. 2005). Some ORs are unique to the larval olfactory organ but others are expressed in both adults and larvae. The extensive characterization of ORN response properties, mapping of OR genes and ORN projection patterns indicates that all information about odours is represented across 40-50 units in the adult and ca. 25 in the larva.

The general rule emerging from these studies is that a single functional class of ORN expresses only one receptor gene and a single receptor gene is expressed only in one class of ORN. One notable exception is the Or83b gene, which is expressed in a large number of ORNs (Vosshall et al. 2000). It was recently shown that in most ORNs this special OR is needed to make the other OR functional (Larsson et al. 2004). Mutations in this gene render the fly largely anosmic. Because Or83b probably does not function as an odour receptor, the dogma of one-neuron-one-receptor is still valid. However, Goldman et al. (2005) have recently demonstrated that pb2B neurons express two 'classical' OR genes (Figure 6). Although both are functional the authors could not show that the response spectrum of the neuron is significantly broadened by this coexpression.

CONCLUDING REMARKS

Chemical ecology of Drosophila

D. melanogaster females lay their eggs on ripe fruit in various stages of fermentation where larvae feed on yeast. Olfaction plays a major role in finding these resources. Simple products of fermentation such as ethanol and acetic acid have long been known to attract *Drosophila* (Barrows 1907; Zhu et al. 2003). Several *Drosophila* ORNs respond to esters, classical components of fruit odours: ethyl acetate, ethyl hexanoate, pentyl acetate, ethyl 3-hydroxybutyrate (De Bruyne et al. 2001; Stensmyr et al. 2003b). However, ovipositing females must not only localize feeding sites with high nutritive value but also with low toxicity. Plant chemical defences could play a role here. Does the presence of a specific receptor for methyl salicylate suggest a role for this compound in the chemical ecology of *Drosophila*? Methyl salicylate is a derivative of salicylic acid and part of volatile distress signals of many plants (Dicke et al. 1990; Shulaev et al. 1997). A predatory mite that has very few ORNs, detects this compound as it is released by the feeding activity of its spider mite prey (De Bruyne et al. 1991; Dicke et al. 1990; De Boer and Dicke 2004). However, ticks also detect it and here the compound is part of an aggregation pheromone emitted when feeding on its mammalian host (Schöni 1987; Diehl et al. 1991), underlining an entirely different role of the same odorant. It is risky to jump to conclusions about the role of specific ORNs in ecological interactions. However, certain odorants may be signals in numerous interactions and their detection a conserved property of many olfactory systems.

Evolution of the olfactory code

Do OR sequences and their associated ORN properties reflect the selective pressure enforced by chemical ecological needs or do they simply vary with phylogenetic distance? Olfactory systems in insects are probably conserved to a certain extent but specific ORNs could be subject to high selective pressure related to shifts in behavioural ecology. The females of several fruit-fly species (Tephritidae) are known to exhibit marked preferences for odours from specific fruits to lay their eggs (Frey and Bush 1990). Although *Drosophila* species are less specific in their selection of oviposition sites, they do show differences in attraction to odours of different fruits (Hoffmann 1985). Within the closely related *melanogaster* species group *D. simulans* shows a lack of preference similar to that of *D. melanogaster*. However, a third member of this group, *D. sechellia*, exhibits remarkable preference for one particular fruit (Higa and Fuyama 1993). A comparison of the response spectra of 8 ORN classes in large basiconic sensilla of the 9 members of this species group indicated that the initial encoding of olfactory information is highly conserved

(Stensmyr et al. 2003a). Only in one ORN class a shift in sensitivity to certain esters was observed in three species, one of them being *D. sechellia*. This species also seems to have replaced one sensillum type with more copies of another. These results show that when changes do occur they can be in the transduction elements themselves (e.g., OR genes) or in genes that regulate the patterning of the antennae. An example of a highly conserved trait is the coexpression of two OR genes, which is also found in *Drosophila pseudoobscura*, a species that diverged ca. 46 million years ago from *D. melanogaster* (Goldman et al. 2005). By contrast, there is only very little sequence homology between OR genes of *D. melanogaster* and a member of the nematoceran Diptera, the mosquito *Anopheles gambiae* (Hill et al. 2002). Detailed knowledge on molecular and cellular elements of the peripheral olfactory system of *Drosophila* is likely to be extremely useful for studying the evolution of odour coding in insects.

ACKNOWLEDGEMENTS

I gratefully acknowledge Kristin Scott, Leslie Vosshall and John Carlson for sending flies carrying GAL4 constructs, and André Fiala, Sören Diegelmann and Erich Buchner for the cameleon flies. I thank Giovanni Galizia for help with calcium imaging and Randolf Menzel for his continued support. The work was financed by the Deutsche Forschungsgemeinschaft (SFB 515-C9).

REFERENCES

Altner, H. and Prillinger, L., 1980. Ultrastructure of invertebrate chemo-, thermo-, and hygrorecepters and its functional significance. *International Review of Cytology,* 67, 69-139.

Ayer, R.K. and Carlson, J., 1992. Olfactory physiology in the *Drosophila* antenna and maxillary palp: acj6 distinguishes two classes of odorant pathways. *Journal of Neurobiology,* 23 (8), 965-982.

Barrows, W.M., 1907. The reactions of the pomace fly, *Drosophila ampelophila* Loew, to odorous substances. *Journal of Experimental Zoology,* 515-537.

Bergmann, A., Agapite, J. and Steller, H., 1998. Mechanisms and control of programmed cell death in invertebrates. *Oncogene,* 17 (25), 3215-3223.

Boekhoff, I., Raming, K. and Breer, H., 1990. Pheromone-induced stimulation of inositol-triphosphate formation in insect antennae is mediated by G-proteins. *Journal of Comparative Physiology. B. Biochemical Systemic and Environmental Physiology,* 160, 99-103.

Brand, A.H. and Perrimon, N., 1993. Targeted gene expression as a means of altering cell fates and generating dominant phenotypes. *Development,* 118 (2), 401-415.

Buck, L. and Axel, R., 1991. A novel multigene family may encode odorant receptors: a molecular basis for odor recognition. *Cell,* 65 (1), 175-187.

Carlson, J.R., 1996. Olfaction in *Drosophila*: from odor to behavior. *Trends in Genetics,* 12 (5), 175-180.

Clyne, P., Grant, A., O'Connell, R., et al. 1997. Odorant response of individual sensilla on the *Drosophila antenna. Invertebrate Neuroscience,* 3 (2/3), 127-135.

Clyne, P.J., Warr, C.G. and Carlson, J.R., 2000. Candidate taste receptors in *Drosophila. Science,* 287 (5459), 1830-1834.

Clyne, P.J., Warr, C.G., Freeman, M.R., et al. 1999. A novel family of divergent seven-transmembrane proteins: candidate odorant receptors in *Drosophila. Neuron,* 22 (2), 327-338.

Davis, R.L., 2004. Olfactory learning. *Neuron,* 44 (1), 31-48.

De Boer, J.G. and Dicke, M., 2004. The role of methyl salicylate in prey searching behavior of the predatory mite *Phytoseiulus persimilis. Journal of Chemical Ecology,* 30 (2), 255-271.

De Bruyne, M., 2003. Physiology and genetics of odor perception in *Drosophila*. *In:* Blomquist, G.J. and Vogt, R.G. eds. *Insect pheromone biochemistry and molecular biology: the biosynthesis and detection of pheonomes and plant volatiles*. Elsevier, New York, 651-697.

De Bruyne, M., Clyne, P.J. and Carlson, J.R., 1999. Odor coding in a model olfactory organ: the *Drosophila* maxillary palp. *Journal of Neuroscience,* 19 (11), 4520-4532.

De Bruyne, M., Dicke, M. and Tjallingii, W.F., 1991. Receptor cell responses in the anterior tarsi of *Phytoseiulus persimilis* to volatile kairomone components. *Experimental and Applied Acarology,* 13 (1), 53-58.

De Bruyne, M., Foster, K. and Carlson, J.R., 2001. Odor coding in the *Drosophila* antenna. *Neuron,* 30 (2), 537-552.

Dicke, M., Van Beek, T.A., Posthumus, M.A., et al. 1990. Isolation and identification of volatile kairomone that affects acarine predator-prey interactions: involvement of host plant in its production. *Journal of Chemical Ecology,* 16 (2), 381-396.

Diehl, P.A., Guerin, P., Vlimant, M., et al. 1991. Biosynthesis, production site, and emission rates of aggregation-attachment pheromone in males of two *Amblyomma* ticks. *Journal of Chemical Ecology,* 17 (5), 833-847.

Dobritsa, A.A., Van der Goes - Van Naters, W., Warr, C.G., et al. 2003. Integrating the molecular and cellular basis of odor coding in the *Drosophila* antenna. *Neuron,* 37 (5), 827-841.

Dunipace, L., Meister, S., McNealy, C., et al. 2001. Spatially restricted expression of candidate taste receptors in the *Drosophila* gustatory system. *Current Biology,* 11 (11), 822-835.

Elmore, T., Ignell, R., Carlson, J.R., et al. 2003. Targeted mutation of a *Drosophila* odor receptor defines receptor requirement in a novel class of sensillum. *Journal of Neuroscience,* 23 (30), 9906-9912.

Fiala, A., Spall, T., Diegelmann, S., et al. 2002. Genetically expressed cameleon in *Drosophila melanogaster* is used to visualize olfactory information in projection neurons. *Current Biology,* 12 (21), 1877-1884.

Frey, J.E. and Bush, G.L., 1990. *Rhagoletis* sibling species and host races differ in host odor recognition. *Entomologia Experimentalis et Applicata,* 57 (2), 123-131.

Galizia, C.G. and Menzel, R., 2001. The role of glomeruli in the neural representation of odours: results from optical recording studies. *Journal of Insect Physiology,* 47 (2), 115-130.

Galizia, C.G. and Vetter, R.S., 2005. Optical methods for analyzing odor-evoked activity in the insect brain. *In:* Christensen, T.A. ed. *Methods in insect neuroscience*. CRC Press, Boca Raton, 349-392.

Gao, Q. and Chess, A., 1999. Identification of candidate *Drosophila* olfactory receptors from genomic DNA sequence. *Genomics,* 60 (1), 31-39.

Gao, Q., Yuan, B.B. and Chess, A., 2000. Convergent projections of *Drosophila* olfactory neurons to specific glomeruli in the antennal lobe. *Nature Neuroscience,* 3 (8), 780-785.

Goldman, A.L., Van der Goes-Van Naters, W., Lessing, D., et al. 2005. Coexpression of two functional odor receptors in one neuron. *Neuron,* 45 (5), 661-666.

Guerin, P.M. and Visser, J.H., 1980. Electroantennogram responses of the carrot fly, *Psila rosae*, to volatile plant components. *Physiological Entomology,* 5 (2), 111-119.

Gupta, B.P. and Rodrigues, V., 1997. Atonal is a proneural gene for a subset of olfactory sense organs in *Drosophila*. *Genes to Cells,* 2 (3), 225-233.

Hallem, E.A., Ho, M.G. and Carlson, J.R., 2004. The molecular basis of odor coding in the *Drosophila* antenna. *Cell,* 117 (7), 965-979.

Higa, I. and Fuyama, Y., 1993. Genetics of food preference in *Drosophila sechellia*. I. Responses to food attractants. *Genetica,* 88 (2/3), 129-136.

Hildebrand, J.G. and Shepherd, G.M., 1997. Mechanisms of olfactory discrimination: converging evidence for common principles across phyla. *Annual Review of Neuroscience,* 20, 595-631.

Hill, C.A., Fox, A.N., Pitts, R.J., et al. 2002. G protein-coupled receptors in *Anopheles gambiae*. *Science,* 298 (5591), 176-178.

Hoffmann, A.A., 1985. Interspecific variation in the response of *Drosophila* to chemicals and fruit odours in a wind tunnel. *Australian Journal of Zoology,* 33 (4), 451-460.

Ito, K., Okada, R., Tanaka, N.K., et al. 2003. Cautionary observations on preparing and interpreting brain images using molecular biology-based staining techniques. *Microscopy Research and Technique,* 62 (2), 170-186.

Jefferis, G.S., Marin, E.C., Watts, R.J., et al. 2002. Development of neuronal connectivity in *Drosophila* antennal lobes and mushroom bodies. *Current Opinion in Neurobiology,* 12 (1), 80-86.

Kaib, M., 1974. Die Fleisch- und Blumenduftrezeptoren auf der Antenne der Schmeissfliege *Calliphora vicina* [The receptors for meat-odour and flower-odour on the antenna of the blowfly *Calliphora vicina*]. *Journal of Comparative Physiology,* 95 (2), 105-121.

Kaissling, K.E., 1995. Single unit and electroantennogram recordings in insect olfactory organs. *In:* Spielman, A.I. and Brand, J.G. eds. *Experimental cell biology of taste and olfaction: current techniques and protocols*. CRC Press, Boca Raton, 361–377.

Kaissling, K.E. and Colbow, K., 1987. *R.H. Wright lectures on insect olfaction*. Simon Fraser University, Burnaby.

Kitamoto, T., 2001. Conditional modification of behavior in *Drosophila* by targeted expression of a temperature-sensitive shibire allele in defined neurons. *Journal of Neurobiology,* 47 (2), 81-92.

Kreher, S.A., Kwon, J.Y. and Carlson, J.R., 2005. The molecular basis of odor coding in the *Drosophila* larva. *Neuron,* 46 (3), 445-456.

Laissue, P.P., Reiter, C., Hiesinger, P.R., et al. 1999. Three-dimensional reconstruction of the antennal lobe in *Drosophila melanogaster*. *Journal of Comparative Neurology,* 405 (4), 543-552.

Larsson, M.C., Domingos, A.I., Jones, W.D., et al. 2004. Or83b encodes a broadly expressed odorant receptor essential for *Drosophila* olfaction. *Neuron,* 43 (5), 703-714.

Lee, T. and Luo, L., 1999. Mosaic analysis with a repressible cell marker for studies of gene function in neuronal morphogenesis. *Neuron,* 22 (3), 451-461.

Löfstedt, C., 1993. Moth pheromone genetics and evolution. *Philosophical Transactions of the Royal Society of London. Series B. Biological Sciences,* 340 (1292), 167-177.

Ng, M., Roorda, R.D., Lima, S.Q., et al. 2002. Transmission of olfactory information between three populations of neurons in the antennal lobe of the fly. *Neuron,* 36 (3), 463-474.

Park, S.K., Shanbhag, S.R., Dubin, A.E., et al. 2002. Inactivation of olfactory sensilla of a single morphological type differentially affects the response of *Drosophila* to odors. *Journal of Neurobiology,* 51 (3), 248-260.

Robertson, H.M., Warr, C.G. and Carlson, J.R., 2003. Molecular evolution of the insect chemoreceptor gene superfamily in *Drosophila melanogaster*. *Proceedings of the National Academy of Sciences of the United States of America,* 100 (Suppl. 2), 14537-14542.

Schneider, D., 1984. Insect olfaction: our research endeavor. *In:* Dawson, J.J. and Enoch, J.M. eds. *Foundations of sensory science*. Springer, Berlin, 381-418.

Schöni, R., 1987. *Das wirtsgebundene Aggregationspheromon der tropischen Buntzecke Amblyomma variegatum Fabricius (Acari: Ixodidae): eine Studie zur Struktur, Wahrnehmung und Wirkungsweise des Pheromons und seiner Komponenten*. PhD Thesis, University of Neuchâtel.

Scott, K., Brady, R., Cravchik, A., et al. 2001. A chemosensory gene family encoding candidate gustatory and olfactory receptors in *Drosophila*. *Cell,* 104 (5), 661-673.

Sengupta, P., Chou, J.H. and Bargmann, C.I., 1996. odr-10 encodes a seven transmembrane domain olfactory receptor required for responses to the odorant diacetyl. *Cell,* 84 (6), 899-909.

Shanbhag, S.R., Hekmat-Scafe, D., Kim, M.S., et al. 2001. Expression mosaic of odorant-binding proteins in *Drosophila* olfactory organs. *Microscopy Research and Technique,* 55 (5), 297-306.

Shulaev, V., Silverman, P. and Raskin, I., 1997. Airborne signalling by methyl salicylate in plant pathogen resistance. *Nature,* 385 (6618), 718-721.

Steinbrecht, R.A., Laue, M. and Ziegelberger, G., 1995. Immunolocalization of pheromone-binding protein and general odorant-binding protein in olfactory sensilla of the silk moths *Antheraea* and *Bombyx*. *Cell and Tissue Research,* 282 (2), 203-217.

Stensmyr, M.C., Dekker, T. and Hansson, B.S., 2003a. Evolution of the olfactory code in the *Drosophila melanogaster* subgroup. *Proceedings of the Royal Society of London. Series B. Biological Sciences,* 270 (1531), 2333-2340.

Stensmyr, M.C., Giordano, E., Balloi, A., et al. 2003b. Novel natural ligands for *Drosophila* olfactory receptor neurones. *Journal of Experimental Biology,* 206 (4), 715-724.

Stocker, R.F., 1994. The organization of the chemosensory system in *Drosophila melanogaster*: a review. *Cell and Tissue Research,* 275 (1), 3-26.

Stocker, R.F., 2001. *Drosophila* as a focus in olfactory research: mapping of olfactory sensilla by fine structure, odor specificity, odorant receptor expression, and central connectivity. *Microscopy Research and Technique,* 55 (5), 284-296.

Störtkuhl, K.F. and Kettler, R., 2001. Functional analysis of an olfactory receptor in *Drosophila melanogaster*. *Proceedings of the National Academy of Sciences of the United States of America,* 98 (16), 9381-9385.

Sweeney, S.T., Broadie, K., Keane, J., et al. 1995. Targeted expression of tetanus toxin light chain in *Drosophila* specifically eliminates synaptic transmission and causes behavioral defects. *Neuron,* 14 (2), 341-351.

Todd, J.L. and Baker, T.C., 1999. Function of peripheral olfactory organs. *In:* Hansson, B.S. ed. *Insect olfaction*. Springer, Berlin, 67-96.

Venkatesh, S., Naresh Singh, R. and Singh, R.N., 1984. Sensilla on the third antennal segment of *Drosophila melanogaster* Meigen (Diptera: Drosophilidae). *International Journal of Insect Morphology and Embryology,* 13 (1), 51-63.

Vosshall, L.B., 2000. Olfaction in *Drosophila*. *Current Opinion in Neurobiology,* 10 (4), 498-503.

Vosshall, L.B., Amrein, H., Morozov, P.S., et al. 1999. A spatial map of olfactory receptor expression in the *Drosophila* antenna. *Cell,* 96 (5), 725-736.

Vosshall, L.B., Wong, A.M. and Axel, R., 2000. An olfactory sensory map in the fly brain. *Cell,* 102 (2), 147-159.

Wang, J.W., Wong, A.M., Flores, J., et al. 2003. Two-photon calcium imaging reveals an odor-evoked map of activity in the fly brain. *Cell,* 112 (2), 271-282.

Warr, C.G., Vosshall, L.B., Amrein, H., et al. 2000. A unified nomenclature system for the *Drosophila* odorant receptors. *Cell,* 102 (2), 145-146.

Wetzel, C.H., Behrendt, H.J., Gisselmann, G., et al. 2001. Functional expression and characterization of a *Drosophila* odorant receptor in a heterologous cell system. *Proceedings of the National Academy of Sciences of the United States of America,* 98 (16), 9377-9380.

Wilson, R.I., Turner, G.C. and Laurent, G., 2004. Transformation of olfactory representations in the *Drosophila* antennal lobe. *Science,* 303 (5656), 366-370.

Xu, P., Atkinson, R., Jones, D.N., et al. 2005. *Drosophila* OBP LUSH is required for activity of pheromone-sensitive neurons. *Neuron,* 45 (2), 193-200.

Zhu, J.W., Park, K.C. and Baker, T.C., 2003. Identification of odors from overripe mango that attract vinegar flies, *Drosophila melanogaster*. *Journal of Chemical Ecology,* 29 (4), 899-909.

CHAPTER 9

CHEMICAL COMMUNICATION BETWEEN ROOTS AND SHOOTS

Towards an integration of aboveground and belowground induced responses in plants

NICOLE M. VAN DAM [#,*] AND T. MARTIJN BEZEMER [#, ##, ###]

[#] *Netherlands Institute of Ecology (NIOO-KNAW), P.O. Box 40, 6666 ZG Heteren, The Netherlands*
[##] *Laboratory of Nematology, Wageningen University and Research Centre, P.O. Box 8123, 6700 ES Wageningen, The Netherlands*
[###] *Laboratory of Entomology, Wageningen University and Research Centre, P.O. Box 8031, 6700 EH Wageningen, The Netherlands*
[*] *Corresponding author. E-mail: n.vandam@nioo.knaw.nl*

Abstract. Induced responses in plants occur in response to both aboveground (AG) and belowground (BG) herbivores and pathogens. So far, the majority of studies have focused on AG induced responses. Possible interactions between AG and BG induced responses have only recently received scientific attention. On the one hand, induction in one plant part may result in systemically induced responses in other parts. On the other hand, simultaneously occurring AG and BG induced responses may interfere, for example, when the activities of root feeders alter the effectiveness of induced responses against leaf feeders. In both cases, AG–BG interactions between induced responses may affect the amount of damage to a plant and therefore constitute an important selection pressure in the evolution of optimal plant-defence strategies.

Here we present a new concept for the integration of AG and BG induced responses in current optimal-defence theory. First, we will consider differences in physiology and morphology between roots and shoots, which relate to their different roles in resource acquisition and which are important in interactions with their environment. Then, we will evaluate how general principles emerging from current theories and mathematical models of optimal AG induced plant defences can be applied to BG induced responses, as well as to their interactions with AG responses. Finally, we argue that plants integrate the information that is communicated by roots and shoots to optimize plant fitness in a multitrophic context.
Keywords: aboveground–belowground interactions; herbivores; inducible defences; nematodes; optimal-defence theory; pathogens; plants; root-induced responses; shoot-induced responses; tolerance

M. Dicke and W. Takken (eds.), Chemical Ecology: From Gene to Ecosystem, 127-143.

INDUCED RESPONSES IN PLANTS

As the main primary producers on this planet, plants serve as food to a large diversity of aboveground and belowground heterotrophic organisms. To protect themselves against this multitude of enemies, plants have evolved a large arsenal of defences, such as trichomes, toxins and digestibility reducers. Many of these defences are inducible, i.e., their production increases when the plant is under attack of herbivores or pathogens. These changes in plants following damage or stress are called 'induced responses' (Karban and Baldwin 1997). They have been found to occur in over 100 plant species and can be elicited by organisms as different in size and feeding strategy as viruses and giraffes (Karban and Baldwin 1997; Agrawal et al. 1999). The compounds or morphological structures, such as trichomes, that are produced in response to an attack may either directly affect the fitness or behaviour of the herbivore, or indirectly affect its survival by attracting or augmenting natural enemies (Vet and Dicke 1992). Generally, induced responses are thought to act as induced defences, i.e., to increase resistance against herbivores and to reduce the negative fitness consequences of herbivory (Karban and Baldwin 1997).

Induced defences are thought to have several advantages over constantly produced constitutive defences. First, it is assumed that induced defences are cost-saving in comparison to constitutive defences, because they are produced only when plants are under attack. When herbivory is absent, the resources that are not used to produce defences may be allocated to growth and reproduction. This is especially beneficial when plants are in competition for limited resources such as light and nutrients (Van Dam and Baldwin 2001). Moreover, high levels of constitutive defences may deter mutualists, such as pollinators and mycorrhizal fungi, which may positively contribute to plant growth and reproduction (Strauss et al. 2002). Inducible defences may allow the plant to decrease defence-compound levels temporarily during mutualistic interactions (Euler and Baldwin 1996). Finally, induced defences are known to be very specific because the plant can obtain 'information' about the herbivore or pathogen that is present before producing defences. Pathogens and herbivores are known to trigger signalling pathways in plants differentially. The plant hormones, jasmonic acid (JA or its methylated form MeJA), salicylic acid (SA or MeSA), ethylene and abscisic acid (ABA), are the best-known compounds for their role in induced responses against insects and other environmental stresses. JA is a product of the lipoxygenase (LOX) signalling pathway that is specifically triggered by herbivore damage (Reymond and Farmer 1998). SA is involved in the signalling pathway that is activated upon pathogen infestation (Hammerschmidt and Smith-Becker 1999; Pieterse et al. 2002). Ethylene and ABA are thought to act mainly as modulators of JA and SA responses, thus enabling the plant to fine-tune its response (Reymond and Farmer 1998; Kahl et al. 2000). This signalling specificity may not only provide information about future risk of herbivory, but also enable the plant to tailor the nature and magnitude of the response to the enemy that is attacking (Karban et al. 1999).

Because defence-related signalling hormones are transported via the vascular system (Zhang and Baldwin 1997) or travel via the air (MeJA, MeSA and ethylene, Kahl et al. 2000; Karban et al. 2004), induced responses are not restricted to the site of attack. In many cases there is a systemically induced response in undamaged plant parts as well. Within shoots, systemic induction patterns generally match source–sink relations and the vascular anatomy of the plant (Davis et al. 1991; Orians et al. 2000; Van Dam et al. 2001). When mature leaves are damaged, undamaged younger – sink – leaves show increased levels of defences as well, whereas undamaged older – source – leaves do not. This may be a functional response of the plant to protect its more valuable photosynthetically active leaves, reflecting an optimal allocation of defence products within the plant (Iwasa et al. 1996; Van Dam et al. 1996; Bezemer et al. 2003).

INTERACTIONS BETWEEN ABOVEGROUND AND BELOWGROUND INDUCED RESPONSES

Although induced plant responses have been studied intensively for over three decades now, induction by belowground (BG) feeding herbivores and how this may affect above ground (AG) herbivores, or AG induced responses and vice-versa, has only recently received scientific attention (Van der Putten et al. 2001; Van Dam et al. 2003). A number of studies have shown that roots employ directly as well as indirectly induced chemical defences against soil pathogens, nematodes and insects (Neori et al. 2000; Van Tol et al. 2001; Walker et al. 2003a; Bauer and Mathesius 2004; Bais et al. 2005). Similar to AG induced responses, the induction by BG herbivores and pathogens may readily result in systemic responses in the leaves. However, with the exception of induced systemic resistance (ISR) and systemic acquired resistance (SAR) by soil bacteria (Pieterse et al. 2002), AG and BG induced responses have rarely been considered in conjunction.

Systemic induction between roots and shoots

Interactions between AG and BG induced defences may occur at different levels. The simplest form of AG–BG interactions is that an induction event in one plant organ alters defence levels in the other organ as well. A review of the literature shows that there are many examples that this may be the case (Table 1). This systemic effect may involve an active up- or down-regulation of genes involved in defence production. Alternatively, the observed changes in defensive chemicals may be a side-effect of reallocation processes after damage. For example, the direction of pyrrolizidine alkaloid induction in artificially shoot-damaged *Cynoglossum officinale* plants appeared to be determined by the genetic strain the plant belonged to. Since the changes in root and shoot alkaloid levels were negatively correlated with each other within half-sib families, the observed changes were assumed to reflect resource reallocation patterns triggered by the damage (Van Dam and Vrieling 1994). Simultaneous reallocation of resources and defence compounds may especially occur when severe artificial damage is applied, which disturbs the shoot–root balance of plants and triggers regrowth responses (Iwasa and Kubo 1997).

Table 1 shows that shoot defence levels may be affected by root-feeding organisms or by root cutting, as well as by decomposers that have no direct

Table 1. *Studies that explicitly measure changes in defence levels in the untreated organ after aboveground or belowground induction. Abbreviations: AG = aboveground, BG = belowground, PI = protease inhibitor, PR protein = pathogen-related protein, suscept. = susceptible genotype, JA = jasmonic acid, SA = salicylic acid*

	Induction by	**Plant**	**Defence**	**Effect**	**Reference**
BG to AG	*Root-invading organisms*				
	Arbuscular mycorrhiza	Plantain	Catalpol	0	Wurst et al. 2004a
		Plantain	Catalpol	+	Gange and West, 1994
	Ectomycorrhiza	Chestnut	Tannins	–	Rieske et al. 2003
		Scots pine	Phenolics	0	Manninen et al. 1998, Manninen et al. 2000
		Scots pine	Terpenes	0	Manninen et al. 1998
	Non-pathogenic bacteria	Arabidopsis	PR gene priming	+	Pieterse et al. 2002
	Plant-feeding nematodes	Black mustard	Glucosinolates	–	Van Dam et al. 2005
		Black mustard	Phenolics	0	Van Dam et al. 2005
		Tobacco resistant	Nicotine	+	Hanounik and Osborne, 1977
		Tobacco suscept.	Nicotine	–	Hanounik and Osborne, 1977
		Rice	Defence genes	+	Blouin et al. 2005
		Potato	PR proteins	+	Rahimi et al. 1996
		Sweet vernal grass	Total phenolics	–	Bezemer et al. 2005
	Root-chewing insects	Black mustard	Glucosinolates	0/+	Van Dam et al. 2005, van Dam et al. unpublished
		Cabbage	Glucosinolates	+	Birch et al. 1992
		Cotton	Terpenoids	+	Bezemer et al. 2004

(cont.)

Table 1 (cont.)

	Induction by	Plant	Defence	Effect	Reference
BG to AG	*Induction mimics*				
	Artificial damage	Ragworth	Alkaloids	0	Hol et al. 2004
		Potato	PI gene	+	Peña-Cortes et al., 1988, Dammann et al. 1997
	JA/SA application	Potato	PI gene	JA +	Dammann et al. 1997
		Black mustard Wild cabbage	Glucosinolates	JA + SA 0/–	Van Dam et al. 2004
		Wild tobacco	PI activity, nicotine	JA +	Baldwin, 1996, van Dam et al. 2001
	Decomposers				
	Earthworms	Plantain	Aucubin	0	Wurst et al. 2004b
		Plantain	Catalpol	–/0	Wurst et al. 2004a, Wurst et al. 2004b
		Plantain	Phytosterols	+	Wurst et al. 2004b
		Rice	Defence genes	+/–	Blouin et al. 2005
AG to BG	*Foliar feeders*				
	Leaf-chewing insects	Cotton	Terpenoids	0	Bezemer et al. 2004
		Ragwort	Alkaloids	–	Hol et al. 2004
	Induction mimics				
	Hormone application	Black mustard Wild cabbage	Glucosinolates	JA 0 SA 0	Van Dam et al. 2004
		Chinese cabbage	Glucosinolates	SA + JA +	Ludwig-Müller et al. 1997
		Okra	PR protein	SA+	Nandi et al. 2003
		Wild tobacco	PI activity	JA +	Van Dam et al. 2001
	Artificial damage	Maize	Hydroxamic acids	0	Collantes et al. 1998
		Rye	Hydroxamic acids	+	Collantes et al. 1999
		Hound's tongue	Alkaloids	+/–	van Dam and Vrieling, 1994

organisms or by root cutting, as well as by decomposers that have no direct interaction with the plant. Based on the data in Table 1, we may conclude that root-chewing insects and application of JA generally increase defence levels in the shoots. This suggests that the JA signalling pathway is involved similarly in the systemic induction from roots to shoots by root-chewing insects, as it is in AG systemic induction by shoot chewers. Even though it has been hypothesized that associations with arbuscular myccorhizal fungi are involved in shoot herbivore specialization (Gange et al. 2002) we found no clear evidence in the literature that this is due to increased levels of defence compounds (Table 1, Wurst et al. 2004a; 2004b). Nematode infestations did not show a clear pattern of changes in shoot defence levels, which may be explained by the different feeding types of the nematode species that were used in the different experiments (Williamson and Gleason 2003).

In contrast to root chewers, neither leaf-chewing insects nor JA application uniformly increased defence levels in roots (Table 1). This suggests that systemic induction from the shoot to the root is not as common as the reverse. A thorough comparison between induction patterns from roots to shoots and the reverse is hampered, however, because we found many more examples of BG induction to affect AG defence levels than the reverse. Possibly, this is due to the practical difficulties involved in quantitatively extracting roots from the soil.

Negative interactions between aboveground and belowground induced responses

AG and BG induced responses may also indirectly affect each other. This may happen when BG and AG herbivores are feeding on the plant at the same time, which is a common situation in natural environments (Van der Putten et al. 2001). As shown above, feeding on one organ may affect defence levels in the other, and when both organs are induced simultaneously, AG and BG induced responses may negatively affect each other. In AG studies it has been shown, for example, that (SA-mediated) pathogen-induced responses may reduce or even inhibit (JA-mediated) herbivore-induced responses (Hammerschmidt and Schultz 1996). Signalling compounds transported from infested roots to the shoot may interact similarly with locally induced hormones triggered by shoot-feeding organisms. In *Brassica nigra* or *B. oleracea* plants, however, we found no evidence that SA application suppresses JA-induced systemic responses when these hormones were applied simultaneously, but spatially separated, to roots and shoots (Van Dam et al. 2004). An experiment that used actual herbivory, however, showed that infestation of *B. nigra* with nematodes or root-fly larvae altered the course of induction in response to shoot-chewing herbivores (Van Dam et al. 2005). Plants increased their shoot defence levels faster when they were infested with nematodes, which suggests that nematodes may prime plants in a way similar to non-pathogenic soil bacteria (Pieterse et al. 2002). Clearly more studies are needed to investigate the generality of this phenomenon.

BG induced responses may also alter 'optimal' defence allocation within the shoot. Cotton plants induced with root-chewing herbivores had a more even distribution of defence compounds among leaves than plants with an AG herbivore (Bezemer et al. 2004). Due to this more even distribution, generalist shoot feeding insects fed less and had reduced growth rates compared to herbivores on plants without root herbivores (Bezemer et al. 2003). Moreover, on cotton plants with root herbivores, extrafloral nectar production was also more evenly distributed among leaves, whereas foliar herbivory caused an increase of extra-floral nectar production specifically for the leaf that was under attack (Wäckers and Bezemer 2003). Because extra-floral nectar serves as an indirect defence by guiding ants to the herbivores, root herbivory thus has the potential to constrain optimal induction of indirect defences in the shoot.

INTEGRATING INTERACTIONS BETWEEN ABOVEGROUND AND BELOWGROUND INDUCED RESPONSES

Both systemically induced responses and negative interactions between simultaneously induced AG and BG induced responses can affect the performance of herbivores and their natural enemies. Consequently, these interactions may affect the amount of damage, and thereby fitness loss, that the plant will suffer. Therefore, interactions between AG and BG induced responses may constitute a significant selection pressure in the evolution of optimal plant-defence strategies. If we want to understand the evolutionary process that has shaped induced responses, BG induced responses must be included (Van der Putten et al. 2001; Van Dam et al. 2003).

In the remainder of this chapter we will present a new conceptual approach to integrate interactions between AG and BG induced responses by focusing on physiological and morphological differences between roots and shoots that are important for their ecological interactions with the environment. Subsequently, we will consider current theories and mathematical models on optimal AG induced plant defences in order to find general principles that may be used to structure new concepts that include BG induced responses.

Differences and similarities between roots and shoots

The differences between roots and shoots in terrestrial plants, of course, are mainly related to the differences in their primary roles in resource acquisition for the plant. Whereas roots primarily acquire water and mineral nutrients from the soil, the primary function of the shoot is to fix carbon via photosynthesis (Hutchings and De Kroon 1994; Taiz and Zeiger 1998). The distinct differences in morphology and physiology of roots and shoots not only reflect the different functions but also the different media in which they forage. The soil in which roots grow is a dense and patchy medium (Crawford et al. 2005). Roots show a high morphological and physiological plasticity in response to the physical and chemical properties of soil. They are able to avoid obstacles, toxins and roots of other plants by guiding the direction of root-tip growth and by controlled withering of tips that grow towards an obstacle (Falik et al. 2005). Moreover, plants can quickly respond to nutrient-rich patches by specifically proliferating into the patch and by increasing local nutrient uptake rates in newly formed root tips (Hutchings and De Kroon 1994; De Kroon et al. 2005). Because the location of nutrient patches in the soil is *a priori* unpredictable, roots forage in many different directions with many different root tips growing simultaneously (Drew 1990).

Other than nutrients in the soil, the distribution of the main AG resource, light, is more homogeneous and unidirectional. Nevertheless, light distribution may also be patchy due to shading by other plants or plant parts. As a consequence, shoots also show several plastic adaptations in response to light availability. In dense populations, for example, shoots are less branched than in open habitats, which is a plastic adaptation to competition with neighbouring plants. Moreover, plants may increase leaf area and reduce leaf thickness when shaded by other plants (Hutchings and De Kroon 1994; De Kroon et al. 2005). As a consequence of the different distributions of AG and BG resources, roots have many more actively growing root tips than shoots have shoot apices, especially in non-clonal herbaceous dicots. Roots also have higher turnover rates than leaves. Damage to a root tip by herbivory or pathogen infection therefore is probably less dramatic for plant growth than the removal of a shoot apical meristem.

Another important difference between roots and shoots is that they grow in environments that are physically very different, which affects the chemical communication with their environment. Roots are constantly and actively excreting a wide array of compounds into the soil, which mainly affect their direct environment, called the rhizophere (Campbell and Greaves 1990; Neori et al. 2000). Root exudation plays a major role in maintaining root–soil contact and in guiding root growth and, thus, in plant survival (Walker et al. 2003a; Bais et al. 2005). The compounds in root exudates may selectively attract and support different micro-organisms that benefit the plant, such as nitrogen-fixing bacteria and mycorrhiza (Walker et al. 2003a). On the other hand, they may also contain defensive compounds that deter pathogenic micro-organisms, fungi and nematodes (Walker et al. 2003b; Bais et al. 2005), or volatile organic compounds that attract natural enemies of root feeders (Van Tol et al. 2001; Rasmann et al. 2005). Root exudates thus may have similar functions as volatile emissions by shoots, for example the attraction of natural enemies of herbivores (Dicke and Van Loon 2000). However, the physical differences between air and soil are responsible for great differences in transport distances and catabolic rates of AG and BG emitted volatiles, for example because UV radiation does not penetrate into the soil (Walker et al. 2003a). There is some evidence that severe artificial shoot damage can increase levels of defensive compounds in root exudates (Collantes et al. 1999). Due to a lack of knowledge on the exact mechanism underlying secretion of phytochemicals by roots, it remains unclear whether this is an active process or a concomitant effect of resource reallocation for regrowth processes (Walker et al. 2003a).

Neither roots nor shoots can survive in isolation but constantly have to exchange their acquired resources as well as coordinate their foraging activities by hormonal signalling (Hutchings and De Kroon 1994). Changes in internal hormone levels also regulate root–shoot regrowth processes after severe damage, especially when the sites of hormone production – root growth tips or shoot apical meristems – are lost (Taiz and Zeiger 1998). Interestingly, the hormones that coordinate root–shoot regrowth after damage, such as auxins, cytokinins, ethylene and ABA, are also involved in modulating induced responses after herbivore damage (Baldwin 1989; Rojo et al. 1998). This emphasizes the importance of considering repair and reallocation processes when studying induced responses.

Optimal plant-defence theory

Central to all theories on optimal defence allocation is that the evolution of plant defences is driven by a cost–benefit balance (Coley et al. 1985; Fagerstrom et al. 1987; Herms and Mattson 1992; Simms 1992; De Jong 1995; Jokela et al. 2000; Shudo and Iwasa 2001; Strauss et al. 2002). In all theories, the benefit is the reduction in damage to the plant, resulting in increased fitness compared to a sub-optimally defended plant. The concept of costs has been debated more intensely. Originally, direct resource investments needed for construction of the defence molecules were considered the principal costs of defence (Gershenzon 1994). However, in many instances, these production costs *per se* were not found to reduce fitness in plants that had higher defence levels than their conspecifics (Bergelson and Purrington 1996). More recently, it has been generally acknowledged that the main costs of defence induction are ecological costs, which occur for example when high defence levels reduce attractiveness to mutualists or competitive strength (Strauss et al. 2002).

Optimal-defence theory also emphasises the value of individual plant parts. If the loss of a certain plant part is reducing plant fitness more than the loss of another plant part, the plant should preferably allocate defence compounds to the former, more valuable part (Van Dam et al. 1995a; Iwasa et al. 1996; Van Dam et al. 2001). The valuation of plant parts has been used as a basis to predict optimal defence allocation as well as optimal defence strategies (Table 2). For example, flowers and seeds, whose survival is highly correlated with plant fitness, often contain very high constitutive levels of defence compounds (Hartmann et al. 1989; Van Dam et al. 1995b; Van Dam et al. 2001). High defence levels are also found in young leaves but, in contrast to flowers, they are still able to increase defence levels after damage (Van Dam et al. 2001; Bezemer et al. 2004). Removing young leaves from a plant significantly reduces future biomass production, whereas removal of old leaves frequently does not (Van Dam et al. 1995b). This again indicates that the high – inducible – defence levels generally found in young leaves reflect optimal defence allocation to more valuable plant parts (McKey 1979; Iwasa et al. 1996).

Several theories include tolerance as an alternative strategy to reduce fitness loss to herbivory (Strauss and Agrawal 1999; Jokela et al. 2000; Fornoni et al. 2004). Originally, defence and tolerance were thought to be mutually exclusive strategies (Van der Meijden et al. 1988), but more recent analyses have revealed that individual plants may use both tolerance and defence to reduce fitness losses (Mauricio et al. 1997; Fornoni et al. 2004).

As for shoots, root parts may differ both in value and vulnerability. Consequently we may expect that different root parts have different optimal strategies when they are damaged (Table 2). Whereas several studies have evaluated the value of different AG plant parts, there is only one study we know of that uses a similar approach for roots (Yanai and Eissenstat 2002). These authors developed a mathematical model and used physiological data on respiration and uptake rates of apple and citrus roots as parameters (Bouma et al. 2001). The model predicted that under high herbivore and pathogen pressure, root life span – and return on investment – could be increased by allocating moderate levels of defences to roots (Yanai and Eissenstat 2002). Under low herbivore pressure, allocation to root defence did not increase root life span. Because this model considers cohorts of roots of the same age, they could not predict differences in defence levels among root parts. However, as stated by the authors, data on the exact costs and benefits of optimal root defence are currently lacking.

Table 2. *Expected values of different shoot and root parts for plant survival and plant fitness and the predicted local optimal defence strategies after pathogen or herbivore damage*

	Damage to plant if lost	**Predicted defence strategy**
Shoot		
Old leaves	–	Tolerance
Young leaves and apical meristem	++	Constitutive and induced defence
Stems	++	Constitutive defence
Flowers/seeds	+++	Constitutive defence
Root		
Tap/main root	++	Constitutive and induced defence
Lateral roots	+	Induced defence
Root tips	–	Tolerance

Another important aspect that is frequently considered in optimal-defence theories, is the likelihood of being attacked. This is especially so for theories that evaluate the costs and benefits of induced vs. constitutive defences. If the likelihood of being attacked is low, induced-defence strategies may be preferred over constitutive defences (Jokela et al. 2000; Shudo and Iwasa 2001). The risk for roots to be attacked by insect herbivores may be much lower than for leaves because roots are less accessible and less nutritious for insects (Hunter 2001). However, roots may have a much higher risk of being attacked by bacteria, fungi and plant-feeding nematodes, with the highest diversity and abundance in the soil (Bongers 1994; Crawford et al. 2005). In response to these abundant root feeders, tolerance may be the preferred strategy (Jokela et al. 2000).

TOWARDS AN INTEGRATION OF ABOVEGROUND AND BELOWGROUND INDUCED RESPONSES IN OPTIMAL-DEFENCE THEORY

The few studies published to date clearly indicate that BG induced defences are important in shaping AG induced responses. In natural environments, plants start to interact with BG organisms as soon as a root has been formed, which usually precedes the onset of shoot emergence (Bezemer and Van Dam in press), and as a consequence, BG induced responses may be very common. Therefore, we argue that BG induced responses must be integrated in optimal plant-defence theory before we can understand the evolution of induced plant responses.

To include AG–BG induced responses in optimal plant-defence theory we need to keep in mind that plants are an integrated system in which roots and shoots *together* contribute to plant fitness. This – trivial, but often ignored – concept also considers the fact that roots and shoots constantly exchange, via hormones, information about their current status and that optimal integration of this information is used to maximize plant fitness in unpredictable AG and BG environments. AG or BG attacks by herbivores or pathogens may affect the status of the roots or the shoots, and as a consequence determine the type of signals that will be produced.

Following attack, plants may first acquire information about the type of organism that is attacking. Based on, for example, bacterial excretions or salivary compounds, plants are able to recognize their attacker (Boland et al. 1995; Mattiacci et al. 1995). After recognition, the plant may produce a specific local signal, for example to initiate localized and rapid death of a few host-plant cells, known as the hypersensitive response, to isolate the site of pathogen infection or oviposition from the rest of the plant (Meiners and Hilker 1997; Hammerschmidt and Nicholson 1999). The locally produced signal or a secondary messenger, however, may also be rapidly transported from the site of damage to other plant parts (Schittko et al. 2000). In some plant species, the systemic signal may simply be required because the site of defence production is remote from the site of damage. Defence compounds that are produced exclusively in the roots, such as nicotine in tobacco or terpenoids in cotton, can only increase in the shoot if there is a systemic signal to trigger defence production in the roots (Zhang and Baldwin 1997). In such plants, the systemic signal results from the physiological organization of the plant species.

Alternatively, the type of the signal may depend on the kind of damage that may be expected. If the attacker is mobile, increases rapidly in population size or is known to spread quickly throughout the whole plant, the plant may benefit by triggering defences in all undamaged plant parts to prepare for the upcoming invasion.

On the other hand, if the organ under attack is damaged to the point at which it will soon lose capacity to acquire its specific resource, it may be more advantageous to signal for reallocation of resources for regrowth and repair (Figure 1). Such a signal may consist of, for example, a decline in auxin production rates after severe damage of the apical shoot meristem (Taiz and Zeiger 1998).

Finally, the plant may be able to compensate for fitness loss after a single attack, but not if another enemy will attack it. In that case, a general systemic defence response may be beneficial to reduce the chance of an additional attack. The latter may be especially beneficial if the plant species has an evolutionary history with several different herbivores that occur sequentially over the growth season (English-Loeb et al. 1993; Viswanathan et al. 2005). The above processes are not mutually exclusive and several signals may be produced at the same time. Possibly there is a

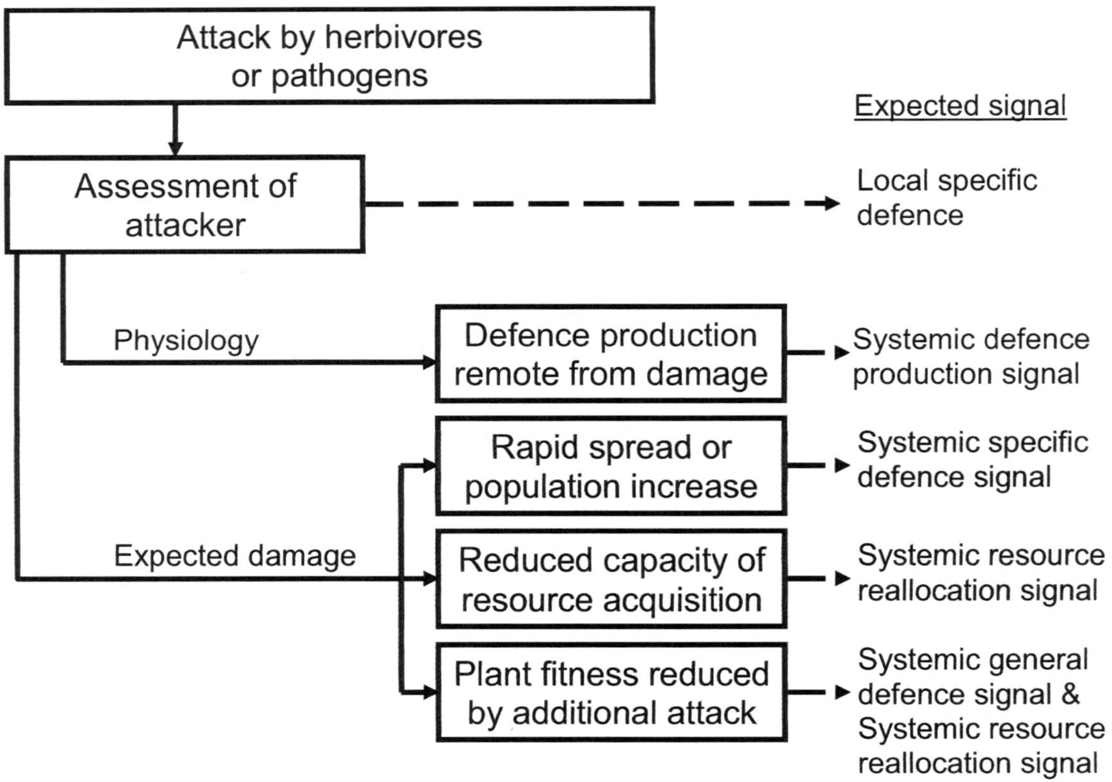

Figure 1. *Conceptual scheme of the types of local and systemic signals that may be produced after an attack by aboveground or belowground herbivores and pathogens. Solid arrows indicate the path of subsequent and parallel processes that may occur in plants that are subject to an attack. Dashed arrows indicate signalling that may result from the process in the box*

hierarchy among these signalling events and the resulting responses, depending on the prevailing environmental conditions in which the plant species has evolved (De Kroon et al. 2005). We therefore cannot speak of a single induced response, but rather of a *suite* of induced responses in roots and shoots that minimizes damage to the plant and optimizes plant fitness as a whole (after Shudo and Iwasa 2001).

It is still unclear how this suite of responses, which may occur in sequence or all at the same time, are integrated to optimize AG and BG induced responses. It may help to consider the temporal aspects of AG and BG induced responses when evaluating the ecological and evolutionary aspects of these interactions (see also Viswanathan et al. 2005). In most natural environments, soil organisms will begin to interact with roots even before the shoots have emerged from the seed coat. The frequency of interactions with AG herbivores and pathogens will increase with shoot size and thus will occur later in time. We therefore argue to focus first on how BG root–soil-organism interactions can affect shoot defence levels and how this can interact with subsequent responses induced by AG feeders (Bezemer and Van Dam in press).

In conclusion, data on interactions between AG and BG induced responses are scarce. More information is especially needed on how these interactions are integrated towards an optimal defence response in plants. In order to raise future experiments above the level of descriptive studies, we need to consider plants as integrated systems and analyse the integration of AG–BG induced responses at different organizational levels, ranging from genes to multitrophic ecological interactions. Only then may we be able to gain a more comprehensive insight into how AG–BG interactions have affected the evolution of induced defences in plants.

ACKNOWLEDGEMENTS

The authors thank Wim H. van der Putten for helpful comments on an earlier version of this manuscript. N.M. van Dam thanks the organizing committee and its chairman Marcel Dicke and for inviting her to participate actively in the Spring School 'Chemical Communication'. N.M. van Dam is supported by a VIDI grant, no. 864-02-001, of the Netherlands Organisation for Scientific Research (NWO) and T.M. Bezemer by a fellowship from the Research School for Production Ecology and Resource Conservation, Wageningen University. Publication 3641 NIOO-KNAW Netherlands Institute of Ecology.

REFERENCES

Agrawal, A.A., Tuzun, S. and Bent, E. (eds.), 1999. *Induced plant defenses against pathogens and herbivores: biochemistry, ecology, and agriculture*. APS Press, St. Paul.

Bais, H.P., Prithiviraj, B., Jha, A.K., et al. 2005. Mediation of pathogen resistance by exudation of antimicrobials from roots. *Nature,* 434 (7030), 217-221.

Baldwin, I.T., 1989. Mechanism of damage-induced alkaloid production in wild tobacco. *Journal of Chemical Ecology,* 15 (5), 1661-1680.

Baldwin, I.T., 1996. Methyl jasmonate-induced nicotine production in *Nicotiana attenuata*: inducing defenses in the field without wounding. *Entomologia Experimentalis et Applicata,* 80 (1), 213-220.

Bauer, W.D. and Mathesius, U., 2004. Plant responses to bacterial quorum sensing signals. *Current Opinion in Plant Biology,* 7 (4), 429-433.

Bergelson, J. and Purrington, C.B., 1996. Surveying patterns in the cost of resistance in plants. *American Naturalist,* 148 (3), 536-558.

Bezemer, T.M., De Deyn, G.B., Bossinga, T.M., et al. 2005. Soil community composition drives aboveground plant-herbivore-parasitoid interactions. *Ecology Letters,* 8 (6), 652-661.

Bezemer, T.M. and Van Dam, N.M., in press. Linking aboveground and belowground induced responses in plants. *Trends in Ecology and Evolution.*

Bezemer, T.M., Wagenaar, R., Van Dam, N.M., et al. 2004. Above- and below-ground terpenoid aldehyde induction in cotton, *Gossypium herbaceum,* following root and leaf injury. *Journal of Chemical Ecology,* 30 (1), 53-67.

Bezemer, T.M., Wagenaar, R., Van Dam, N.M., et al. 2003. Interactions between above- and belowground insect herbivores as mediated by the plant defense system. *Oikos,* 101 (3), 555-562.

Birch, A.N.E., Griffiths, D.W., Hopkins, R.J., et al. 1992. Glucosinolate responses of swede, kale, forage and oilseed rape to root damage by turnip root fly (*Delia floralis*) larvae. *Journal of the Science of Food and Agriculture,* 60 (1), 1-9.

Blouin, M., Zuily-Fodil, Y., Pham-Thi, A.T., et al. 2005. Belowground organism activities affect plant aboveground phenotype, inducing plant tolerance to parasites. *Ecology Letters,* 8 (2), 202-208.

Boland, W., Hopke, J., Donath, J., et al. 1995. Jasmonic acid and coronatin induced odor production in plants. *Angewandte Chemie,* 34 (15), 1600-1602.

Bongers, T., 1994. *De nematoden van Nederland: een identificatietabel voor de in Nederland aangetroffen zoetwater- en bodembewonende nematoden.* 2nd edn. Koninklijke Nederlandse Natuurhistorische Vereniging, Utrecht. Natuurhistorische bibliotheek van de KNNV no. 46.

Bouma, T.J., Yanai, R.D., Elkin, A.D., et al. 2001. Estimating age-dependent costs and benefits of roots with contrasting life span: comparing apples and oranges. *New Phytologist,* 150 (3), 685-695.

Campbell, R. and Greaves, M.P., 1990. Anatomy and community structure of the rhizosphere. *In:* Lynch, J.M. ed. *The rhizosphere.* Wiley, Chichester, 11-34.

Coley, P.D., Bryant, J.P. and Chapin, F.S., 1985. Resource availability and plant antiherbivore defense. *Science,* 230 (4728), 895-899.

Collantes, H.G., Gianoli, E. and Niemeyer, H.M., 1998. Changes in growth and chemical defences upon defoliation in maize. *Phytochemistry,* 49 (7), 1921-1923.

Collantes, H.G., Gianoli, E. and Niemeyer, H.M., 1999. Defoliation affects chemical defenses in all plant parts of rye seedlings. *Journal of Chemical Ecology,* 25 (3), 491-499.

Crawford, J.W., Harris, J.A., Ritz, K., et al. 2005. Towards an evolutionary ecology of life in soil. *Trends in Ecology and Evolution,* 20 (2), 81-87.

Dammann, C., Rojo, E. and Sanchez-Serrano, J.J., 1997. Abscisic acid and jasmonic acid activate wound-inducible genes in potato through separate, organ-specific signal transduction pathways. *Plant Journal,* 11 (4), 773-782.

Davis, J.M., Gordon, M.P. and Smit, B.A., 1991. Assimilate movement dictates remote sites of wound-induced gene expression in poplar leaves. *Proceedings of the National Academy of Sciences of the United States of America,* 88 (6), 2393-2396.

De Jong, T.J., 1995. Why fast-growing plants do not bother about defence. *Oikos,* 74 (3), 545-548.

De Kroon, H., Huber, H., Stuefer, J.F., et al. 2005. A modular concept of phenotypic plasticity in plants. *New Phytologist,* 166 (1), 73-82.

Dicke, M. and Van Loon, J.J.A., 2000. Multitrophic effects of herbivore-induced plant volatiles in an evolutionary context. *Entomologia Experimentalis et Applicata,* 97 (3), 237-249.

Drew, M.C., 1990. Root function, development, growth and mineral nutrition. *In:* Lynch, J.M. ed. *The rhizosphere.* Wiley, Chichester, 35-57.

English-Loeb, G.M., Karban, R. and Hougen-Eitzman, D., 1993. Direct and indirect competition between spider mites feeding on grapes. *Ecological Applications,* 3 (4), 699-707.

Euler, M. and Baldwin, I.T., 1996. The chemistry of defense and apparency in the corollas of *Nicotiana attenuata. Oecologia,* 107 (1), 102-112.

Fagerstrom, T., Larsson, S. and Tenow, O., 1987. On optimal defense in plants. *Functional Ecology,* 1 (2), 73-82.

Falik, O., Reides, P., Gersani, M., et al. 2005. Root navigation by self inhibition. *Plant Cell and Environment,* 28 (4), 562-569.

Fornoni, J., Nunez-Farfan, J., Valverde, P.L., et al. 2004. Evolution of mixed strategies of plant defense allocation against natural enemies. *Evolution,* 58 (8), 1685-1695.

Gange, A.C., Stagg, P.G. and Ward, L.K., 2002. Arbuscular mycorrhizal fungi affect phytophagous insect specialism. *Ecology Letters,* 5 (1), 11-15.

Gange, A.C. and West, H.M., 1994. Interactions between arbuscular mycorrhizal fungi and foliar-feeding insects in *Plantago lanceolata* L. *New Phytologist,* 128 (1), 79-87.

Gershenzon, J., 1994. Metabolic costs of terpenoid accumulation in higher plants. *Journal of Chemical Ecology,* 20 (6), 1281-1328.

Hammerschmidt, R. and Nicholson, R.L., 1999. A survey of plant defense responses to pathogens. *In:* Agrawal, A.A., Tuzun, S. and Bent, E. eds. *Induced plant defenses against pathogens and herbivores.* APS Press, St. Paul, 55-72.

Hammerschmidt, R. and Schultz, J.C., 1996. Multiple defenses and signals in plant defense against pathogens and herbivores. *In:* Romeo, J.T., Saunders, J.A. and Barbosa, P. eds. *Phytochemical diversity and redundancy in ecological interactions: [proceedings of the thirty-fifth annual meeting of the Phytochemical Society of North America., held August 12 - 16, 1995, in Sault Ste. Marie, Ontario, Canada].* Plenum Press, New York, 121-155. Recent Advances in Phytochemistry no. 30.

Hammerschmidt, R. and Smith-Becker, J.A., 1999. The role of salicylic acid in disease resistance. *In:* Agrawal, A.A., Tuzun, S. and Bent, E. eds. *Induced plant defenses against pathogens and herbivores.* APS Press, St. Paul, 37-53.

Hanounik, S.B. and Osborne, W.W., 1977. The relationships between population density of *Meloidogyne incognita* and nicotine content of tobacco. *Nematologica,* 23 (2), 147-152.

Hartmann, T., Ehmke, A., Eilert, U., et al. 1989. Sites of synthesis, translocation and accumulation of pyrrolizidine alkaloid n-oxides in *Senecio vulgaris* L. *Planta,* 177 (1), 98-107.

Herms, D.A. and Mattson, W.J., 1992. The dilemma of plants: to grow or defend. *Quarterly Review of Biology,* 67 (3), 283-335.

Hol, W.H.G., Macel, M., Van Veen, J.A., et al. 2004. Root damage and aboveground herbivory change concentration and composition of pyrrolizidine alkaloids of *Senecio jacobaea. Basic and Applied Ecology,* 5 (3), 253-260.

Hunter, M.D., 2001. Out of sight, out of mind: the impacts of root-feeding insects in natural and managed systems. *Agricultural and Forest Entomology,* 3 (1), 3-9.

Hutchings, M.J. and De Kroon, H., 1994. Foraging in plants: the role of morphological plasticity in resource acquisition. *Advances in Ecological Research,* 25, 159-238.

Iwasa, Y. and Kubo, T., 1997. Optimal size of storage for recovery after unpredictable disturbances. *Evolutionary Ecology,* 11 (1), 41-65.

Iwasa, Y., Kubo, T., Van Dam, N., et al. 1996. Optimal level of chemical defense decreasing with leaf age. *Theoretical Population Biology,* 50 (2), 124-148.

Jokela, J., Schmid-Hempel, P. and Rigby, M.C., 2000. Dr. Pangloss restrained by the Red Queen: steps towards a unified defence theory. *Oikos,* 89 (2), 267-274.

Kahl, J., Siemens, D.H., Aerts, R.J., et al. 2000. Herbivore-induced ethylene suppresses a direct defense but not a putative indirect defense against an adapted herbivore. *Planta,* 210 (2), 336-342.

Karban, R., Agrawal, A.A., Thaler, J.S., et al. 1999. Induced plant responses and information content about risk of herbivory. *Trends in Ecology and Evolution,* 14 (11), 443-447.

Karban, R. and Baldwin, I.T., 1997. *Induced responses to herbivory.* University of Chicago Press, Chicago.

Karban, R., Huntzinger, M. and McCall, A.C., 2004. The specificity of eavesdropping on sagebrush by other plants. *Ecology,* 85 (7), 1846-1852.

Ludwig-Müller, J., Schubert, B., Pieper, K., et al. 1997. Glucosinolate content in susceptible and resistant Chinese cabbage varieties during development of clubroot disease. *Phytochemistry,* 44 (3), 407-414.

Manninen, A.M., Holopainen, T. and Holopainen, J.K., 1998. Susceptibility of ectomycorrhizal and non-mycorrhizal Scots pine (*Pinus sylvestris*) seedlings to a generalist insect herbivore, *Lygus rugulipennis*, at two nitrogen availability levels. *New Phytologist,* 140 (1), 55-63.

Manninen, A.M., Holopainen, T., Lyytikainen-Saarenmaa, P., et al. 2000. The role of low-level ozone exposure and mycorrhizas in chemical quality and insect herbivore performance on Scots pine seedlings. *Global Change Biology,* 6 (1), 111-121.

Mattiacci, L., Dicke, M. and Posthumus, M.A., 1995. Beta-glucosidase: an elicitor of herbivore-induced plant odor that attracts host-searching parasitic wasps. *Proceedings of the National Academy of Sciences of the United States of America,* 92 (6), 2036-2040.

Mauricio, R., Rausher, M.D. and Burdick, D.S., 1997. Variation in the defense strategies of plants: are resistance and tolerance mutually exclusive? *Ecology,* 78 (5), 1301-1311.

McKey, D., 1979. The ditribution of secondary plant metabolites within plants. *In:* Rosenthal, G.A. and Janzen, D.H. eds. *Herbivores: their interaction with secondary plant metabolites.* Academic Press, New York, 56-134.

Meiners, T. and Hilker, M., 1997. Host location in *Oomyzus gallerucae* (Hymenoptera: Eulophidae), an egg parasitoid of the elm leaf beetle *Xanthogaleruca luteola* (Coleoptera: Chrysomelidae). *Oecologia,* 112 (1), 87-93.

Nandi, B., Kundu, K., Banerjee, N., et al. 2003. Salicylic acid-induced suppression of *Meloidogyne incognita* infestation of okra and cowpea. *Nematology,* 5 (5), 747-752.

Neori, A., Reddy, K.R., Ciskova-Koncalova, H., et al. 2000. Bioactive chemicals and biological-biochemical activities and their functions in rhizospheres of wetland plants. *Botanical Review,* 66 (3), 350-378.

Orians, C.M., Pomerleau, J. and Ricco, R., 2000. Vascular architecture generates fine scale variation in systemic induction of proteinase inhibitors in tomato. *Journal of Chemical Ecology,* 26 (2), 471-485.

Peña-Cortes, H., Sanchez-Serrano, J., Rocha-Sosa, M., et al. 1988. Systemic induction of proteinase inhibitor-Ii gene expression in potato plants by wounding. *Planta,* 174 (1), 84-89.

Pieterse, C.M.J., Van Wees, S.C.M., Ton, J., et al. 2002. Signalling in rhizobacteria-induced systemic resistance in *Arabidopsis thaliana. Plant Biology,* 4 (5), 535-544.

Rahimi, S., Perry, R.N. and Wright, D.J., 1996. Identification of pathogenesis-related proteins induced in leaves of potato plants infected with potato cyst nematodes, *Globodera* species. *Physiological and Molecular Plant Pathology,* 49 (1), 49-59.

Rasmann, S., Kollner, T.G., Degenhardt, J., et al. 2005. Recruitment of entomopathogenic nematodes by insect-damaged maize roots. *Nature,* 434 (7034), 732-737.

Reymond, P. and Farmer, E.E., 1998. Jasmonate and salicylate as global signals for defense gene expression. *Current Opinion in Plant Biology,* 1 (5), 404-411.

Rieske, L.K., Rhoades, C.C. and Miller, S.P., 2003. Foliar chemistry and gypsy moth, *Lymantria dispar* (L.), herbivory on pure American chestnut, *Castanea dentata* (Fam: Fagaceae), and a disease-resistant hybrid. *Environmental Entomology,* 32 (2), 359-365.

Rojo, E., Titarenko, E., Leon, J., et al. 1998. Reversible protein phosphorylation regulates jasmonic acid-dependent and -independent wound signal transduction pathways in *Arabidopsis thaliana. Plant Journal,* 13 (2), 153-165.

Schittko, U., Preston, C.A. and Baldwin, I.T., 2000. Eating the evidence? *Manduca sexta* larvae can not disrupt specific jasmonate induction in *Nicotiana attenuata* by rapid consumption. *Planta,* 210 (2), 343-346.

Shudo, E. and Iwasa, Y., 2001. Inducible defense against pathogens and parasites: optimal choice among multiple options. *Journal of Theoretical Biology,* 209 (2), 233-247.

Simms, E.L., 1992. Costs of plant resistance to herbivory. *In:* Fritz, R.S. and Simms, E.L. eds. *Plant resistance to herbivores and pathogens: ecology, evolution, and genetics.* University of Chicago Press, Chicago, 392-425.

Strauss, S.Y. and Agrawal, A.A., 1999. The ecology and evolution of plant tolerance to herbivory. *Trends in Ecology and Evolution,* 14 (5), 179-185.

Strauss, S.Y., Rudgers, J.A., Lau, J.A., et al. 2002. Direct and ecological costs of resistance to herbivory. *Trends in Ecology and Evolution,* 17 (6), 278-285.

Taiz, L. and Zeiger, E., 1998. *Plant physiology.* 2nd edn. Sinauer, Sunderland.

Van Dam, N.M., Harvey, J.A., Wäckers, F.L., et al. 2003. Interactions between aboveground and belowground induced responses against phytophages. *Basic and Applied Ecology,* 4 (1), 63-77.

Van Dam, N.M. and Baldwin, I.T., 2001. Competition mediates costs of jasmonate-induced defences, nitrogen acquisition and transgenerational plasticity in *Nicotiana attenuata. Functional Ecology,* 15 (3), 406-415.

Van Dam, N.M., De Jong, T.J., Iwasa, Y., et al. 1996. Optimal distribution of defences: are plants smart investors? *Functional Ecology,* 10 (1), 128-136.

Van Dam, N.M., Horn, M., Mareš, M., et al. 2001. Ontogeny constrains systemic protease inhibitor response in *Nicotiana attenuata. Journal of Chemical Ecology,* 27 (3), 547-568.

Van Dam, N.M., Raaijmakers, C.E. and Van der Putten, W.H., 2005. Root herbivory reduces growth and survival of the shoot feeding specialist *Pieris rapae* on *Brassica nigra. Entomologia Experimentalis et Applicata,* 115 (1), 161-170.

Van Dam, N.M. and Vrieling, K., 1994. Genetic variation in constitutive and inducible pyrrolizidine alkaloid levels in *Cynoglossum officinale* L. *Oecologia,* 99 (3/4), 374-378.
Van Dam, N.M., Vuister, L.W.M., Bergshoeff, C., et al. 1995a. The "raison d'etre" of pyrrolizidine alkaloids in *Cynoglossum officinale*: deterrent effects against generalist herbivores. *Journal of Chemical Ecology,* 21 (5), 507-523.
Van Dam, N.M., Witjes, L. and Svatoš, A., 2004. Interactions between aboveground and belowground induction of glucosinolates in two wild *Brassica* species. *New Phytologist,* 161 (3), 801-810.
Van Dam, N.M., Witte, L., Theuring, C., et al. 1995b. Distribution, biosynthesis and turnover of pyrrolizidine alkaloids in *Cynoglossum officinale. Phytochemistry,* 39 (2), 287-292.
Van der Meijden, E., Wijn, M. and Verkaar, H.J., 1988. Defense and regrowth alternative plant strategies in the struggle against herbivores. *Oikos,* 51 (3), 355-363.
Van der Putten, W.H., Vet, L.E.M., Harvey, J.A., et al. 2001. Linking above- and belowground multitrophic interactions of plants, herbivores, pathogens, and their antagonists. *Trends in Ecology and Evolution,* 16 (10), 547-554.
Van Tol, R.W.H.M., Van der Sommen, A.T.C., Boff, M.I.C., et al. 2001. Plants protect their roots by alerting the enemies of grubs. *Ecology Letters,* 4 (4), 292-294.
Vet, L.E.M. and Dicke, M., 1992. Ecology of infochemical use by natural enemies in a tritrophic context. *Annual Review of Entomology,* 37 (1), 141-172.
Viswanathan, D.V., Narwani, A.J.T. and Thaler, J.S., 2005. Specificity in induced plant responses shapes patterns of herbivore occurrence on *Solanum dulcamara. Ecology,* 86 (4), 886-896.
Wäckers, F.L. and Bezemer, T.M., 2003. Root herbivory induces an above-ground indirect defence. *Ecology Letters,* 6 (1), 9-12.
Walker, T.S., Bais, H.P., Grotewold, E., et al. 2003a. Root exudation and rhizosphere biology. *Plant Physiology,* 132 (1), 44-51.
Walker, T.S., Bais, H.P., Halligan, K.M., et al. 2003b. Metabolic profiling of root exudates of Arabidopsis thaliana. *Journal Of Agricultural And Food Chemistry,* 51 (9), 2548-2554.
Williamson, V.M. and Gleason, C.A., 2003. Plant-nematode interactions. *Current Opinion in Plant Biology,* 6 (4), 327-333.
Wurst, S., Dugassa-Gobena, D., Langel, R., et al. 2004a. Combined effects of earthworms and vesicular-arbuscular mycorrhizas on plant and aphid performance. *New Phytologist,* 163 (1), 169-176.
Wurst, S., Dugassa-Gobena, D. and Scheu, S., 2004b. Earthworms and litter distribution affect plant-defensive chemistry. *Journal of Chemical Ecology,* 30 (4), 691-701.
Yanai, R.D. and Eissenstat, D.M., 2002. Coping with herbivores and pathogens: a model of optimal root turnover. *Functional Ecology,* 16 (6), 865-869.
Zhang, Z.P. and Baldwin, I.T., 1997. Transport of (2-^{14}C)jasmonic acid from leaves to roots mimics wound-induced changes in endogenous jasmonic acid pools in *Nicotiana sylvestris. Planta,* 203 (4), 436-441.

CHAPTER 10

FOOD-WEB INTERACTIONS IN LAKES

What is the impact of chemical information conveyance?

ELLEN VAN DONK

NIOO-KNAW Centre for Limnology, Rijksstraatweg 6, 3631 AC Nieuwersluis, The Netherlands. E-mail: e.vandonk@nioo.knaw.nl

Abstract. The structure of aquatic ecosystems is determined by complex interactions among individual organisms at different trophic levels. Although our basic understanding of how top-down and bottom-up processes interact to determine food-web dynamics has advanced, we still lack insights into how complex interactions and feedbacks affect the dynamics and structure of food webs. It is now becoming increasingly clear that, in addition to energy transfer from one trophic level to the other, there is exchange of information between these levels, facilitated by the release of infochemicals by the organisms. There is evidence from recent studies that the exchange of chemical information in freshwater ecosystems is likely to play a decisive role in shaping structure and functioning of these systems. Chemical communication among freshwater organisms mediates many aspects of both predation and interspecific competition, which play key roles in determining the community structure and ecosystem functioning. For example, consumer-induced defences in phytoplankton and zooplankton include modifications in the characteristics relating to life history, behaviour, morphology and biochemistry. These inducible defences affect trophic interactions by altering predator feeding rates through changes in attack rate or handling time or both. Also host-specific fungal parasitism in phytoplankton is probably controlled by infochemicals. The motile fungi recognize their host by host-secreted compounds. In this chapter I will discuss how infochemicals may affect the dynamics and structure of planktonic food webs.
Keywords: induced defence; phenotypic plasticity; infochemicals; plankton; population dynamics; ecosystem effects

INTRODUCTION

Among terrestrial organisms we consider it self-evident that interactions are not only influenced by visual signals but also by chemical signals, for example in predator–prey interactions. In aquatic systems, however, interactions based on chemical information transfer are less obvious (Brönmark and Hansson 2000). Predation is an important mortality factor for planktonic species; therefore, many planktonic organisms have developed a wide variety of defences to avoid predation by higher trophic levels. Many phytoplankton species are notoriously flexible in their morphology, growth form and biochemical composition. For example, several of these variable traits in phytoplankton have been interpreted as defence mechanisms against grazing. Pelagic phytoplankton employs different defence strategies to avoid

M. Dicke and W. Takken (eds.), Chemical Ecology: From Gene to Ecosystem, 145-160.

being ingested and, if ingested, to pass unharmed through the grazer's gut. Zooplankton feed with differing success on various phytoplankton species, depending primarily on size, shape, cell-wall structure and the production of toxins. Some evidence for size-related effects comes from experiments that involve feeding zooplankton with particles of different sizes, keeping their shape constant. For example, Burns (1968) found a clear relationship between the grazer's body size and the maximum size of spherical beads that can be ingested. Hardness of algae also influences ingestibility (DeMott 1995). Gelatinous chlorophytes may be readily ingested but are poorly digested by zooplankters like *Daphnia* (Porter 1975), resulting in depressed zooplankton growth rates (Stutzman 1995). Zooplankton rarely feed on filamentous cyanobacteria because they are large and can be toxic (Lampert 1987). Further, extracellular substances released from cyanobacteria inhibit the grazing activity of daphnids (Haney et al. 1994). In contrast, detritus generated from filamentous cyanobacteria is both better ingested and assimilated by *Daphnia* spp. (Gulati et al. 2001). Mucus excretion by diatoms also inhibits copepod grazing (Malej and Harris 1993). Finally, nutrient-deficient algae may also be grazed with decreased efficiency, owing to either reduced ingestion rates (Sterner and Smith 1993) or reduced assimilation efficiency (Van Donk and Hessen 1993; Van Donk et al. 1997), which increases the probability of persistence of such algae during periods of low growth rates.

Some of these changes in defensive traits in the field can be explained by clonal replacement as conditions change (Wood and Leatham 1992; Yoshida et al. 2003). However, there is also evidence for phenotypic plasticity. For example, the dinoflagellate *Ceratium* shows considerable phenotypic plasticity in its horn lengths (Hutchinson 1967). On the other hand, the cyanobacterium *Microcystis* may phenotypically vary in its toxic effects (Benndorf and Henning 1989). The green algal genus *Scenedesmus* is notoriously phenotypically flexible (Trainor and Egan 1991). Individual strains of various *Scenedesmus* species can grow as unicells or form colonies (coenobia) of four or eight cells. The cells can also vary in the number and size of the spines.

It is well known that many algal species isolated as clones from the field change their morphology or growth form after several generations in laboratory cultures, suggesting that some unknown factor triggers their 'typical' or consistent appearance in the field. For example, spiny algae like *Staurastrum* lose their bizarre form, colonies like *Microcystis* grow as single cells, and flakes of *Aphanizomenon* grow as single filaments. In the field, large flakes of *Aphanizomenon* are frequently found in the presence of large *Daphnia* (Lynch 1980). A similar phenomenon has been observed in the diatom *Synedra*, which occurs as colonies consisting of dozens to hundreds of cells when *Daphnia* is present, but as single cells in the grazer's absence. It is, however, difficult to determine if the observed effect is caused by selective grazing on small flakes and single cells or an active response to the grazers' presence.

Recently, several studies have shown that not only in terrestrial but also in aquatic systems many organisms are receptive to chemical signals exuded not only by conspecifics but also by potential predators and grazers, which help them gather information about their environment. This so-called chemical communication is mediated by information-conveying chemicals (infochemicals) and is a well-known ecological phenomenon that facilitates interactions between organisms (Dicke and Sabelis 1992). Infochemicals are defined as chemical compounds that convey information between individuals and thereby evoke a physiological or behavioural response in the receiver (Dicke and Sabelis 1988). Consumer-induced defences in phytoplankton and zooplankton include changes in morphology (e.g., formation of spines and colonies), biochemistry (e.g., production of toxins, repellents), behavioural responses (e.g., migration, refuge use) and in life-history characteristics. Below I will review consumer-induced defences in plankton and discuss benefits and costs of induced defences and their impact on population dynamics and ecosystem functioning. Furthermore, I will discuss the current knowledge of the chemical nature of aquatic infochemicals and their transportation in the water. I also shortly review allelopathic interactions, attraction to food by means of infochemicals, and multitrophic indirect defences.

CONSUMER-INDUCED DEFENCES IN PHYTOPLANKTON

Hessen and Van Donk (1993) discovered that a chemical released from grazing *Daphnia* induced the formation of colonies in the green alga *Scenedesmus subspicatus*. On exposure to water in which daphnids had been cultured, *Scenedesmus* formed numerous large, four- to eight-celled colonies, with more rigid and longer spines (Figure 1). The induced changes in the algae conferred grazing resistance against small zooplankters and can be interpreted as an adaptive anti-herbivore strategy. Reduced algal palatability adversely affected feeding rates in zooplankton and reduced their growth rates and fecundity (Van Donk et al. 1999). Lampert et al. (1994) confirmed the findings that colony formation was mediated by chemicals released by daphnids, by adding water in which *Daphnia* had been swimming to spineless *Scenedesmus acutus*. Van Donk et al. (1999) examined the effect of *Daphnia* infochemicals on the morphology of fifteen strains of Chlorophyceae, two strains of Bacillariophyceae and three strains of Cyanophyceae. *Daphnia*-induced colony formation, which was restricted to Chlorophyceae was, in addition to the genus *Scenedesmus,* also observed in *Coelastrum*. Verschoor et al. (2004a) showed that species of Scenedesmaceae that responded to *Daphnia,* generally also responded to infochemicals from the rotifer *Brachionus calyciflorus*, and that colony size could be related to infochemical concentration.

The colony formation in response to grazing-associated infochemicals does not seem to be unique to freshwater algae: it has been reported by Wolfe et al. (1997) for marine phytoplankton (*Phaeocystis*, a haptophyte) in response to infochemicals released by zooplankton.

In the desmid *Staurastrum* the presence of *Daphnia* induced the formation of mucus and clumping of the algal cells, making them less edible for zooplankton. However, this was caused by the stirring of the water due to filtering activity of daphnids and not to an infochemical (Wiltshire et al. 2003).

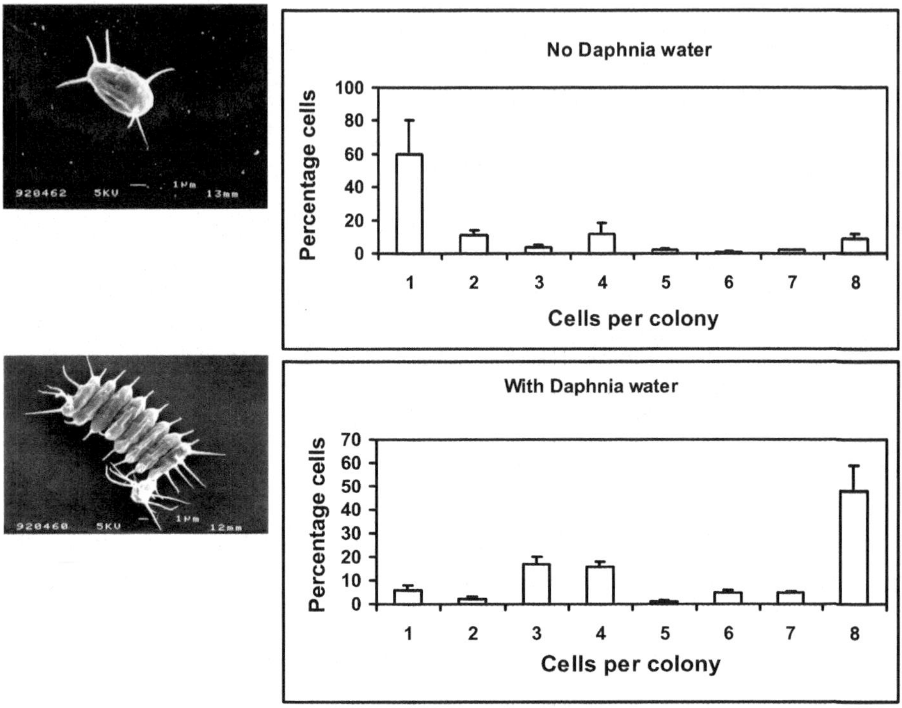

Figure 1. *Colony induction in a Scenedesmus culture in response to an infochemical released by Daphnia magna. Percentage of total number of cells forming colonies of varying cell numbers after 48 hours of growth. Single cells dominate in the cultures without Daphnia water (upper panel), while eight-cell colonies dominate in the infochemical treatments (lower panel) (From Van Donk et al. 1999)*

Jang et al. (2003) demonstrated that several strains of *Microcystis aeruginosa* increased toxin production in response to direct and indirect exposure to herbivorous zooplankton. This supports the hypothesis that this response is an induced defence strategy, mediated by the release of infochemicals from zooplankton.

Hansson (1996; 2000) reported that several freshwater algal species may possibly regulate their recruitment rate from sediment depending on the presence or absence of grazers in the water column. For example, flagellated algae like *Gonyostomum semen* can use infochemicals released by herbivores to adjust the timing of their recruitment from the 'seed-bank', thereby reducing the exposure to grazing.

CONSUMER-INDUCED DEFENCES IN ZOOPLANKTON

Predator-induced responses in zooplankton have elicited increasing research interest during the last two decades. Since Larsson and Dodson (1993) reviewed the state-of-the-art research on chemical communication in planktonic animals, several studies on such aspects and new research lines on anti-predator defences in zooplankton have developed (Tollrian and Harvell 1999; Brönmark and Hansson 2000; Lass and Spaak 2003).

Rotifers, viz., *Brachionus calyciflorus*, have been the first planktonic organisms for which responses to infochemicals derived from their predators were observed (Gilbert 1966). Inducible morphological changes have also been described for ciliates of the genus *Euplotes* upon their exposure to infochemicals from ciliate predators of the genus *Lembadion* (Kuhlmann and Heckmann 1985).

The cladoceran *Daphnia* has developed several behavioural defence mechanisms. The daphnids are not only a good food source for planktivorous fish but also for invertebrate predators, such as *Chaoborus* larvae and *Notonecta*. One of the frequently studied behavioural response mechanisms in daphnids is diel vertical migration (DVM). *Daphnia* migrates down during the day to relatively darker, deeper waters to avoid fish. During the night they ascend to the warmer surface water layers of the lake, where algal food is more abundant. Although light plays an important role, DVM seems to be triggered by a chemical signal (Dodson 1988).

An analogous mechanism to DVM is diel horizontal migration (DHM) of *Daphnia* into the vicinity of water plants. The macrophytes act as shelter for the daphnids to protect them from fish during the day (Burks et al. 2000). Interestingly, these macrophytes have been shown to produce chemicals that repel *Daphnia*, and only the infochemicals exuded by fish can override the repellent effect of the chemical compounds produced by macrophytes (Burks et al. 2000).

Furthermore, *Daphnia* can change morphologically following exposure to predator infochemicals. These changes occur to a certain degree as a response to seasonal temperature changes, but also in response to infochemicals. There are reports of the formation of neckteeth, helmets and crests in *Daphnia* exposed to, respectively, *Chaoborus* (Krueger and Dodson 1981), planktivorous fish (Tollrian 1994) and a notonectid predator (Grant and Bayly 1981). Apart from this, Slusarczyk (1999) reported evidence for two chemical cues that regulate synchronization of sexual reproduction (formation of males and production of ephippial eggs or resting eggs) to protect the genome during periods of high predation risk. Not only does *Daphnia* react to infochemicals released from predators, but also to a chemical signal released by injured conspecifics. Filtered water that had contained crushed daphnids before filtration induced individual daphnids to remain deeper in the water column. They also aggregated more frequently in the presence of a chemical released by fish (Pijanowska and Kowalczewski 1997). This may be the result of mechanical interference in food collection and allelochemical interactions.

Zooplankters can also defend themselves by becoming less visible to their predators. Copepods in arctic, oligotrophic lakes are often pigmented by carotenoids, which protect them against UV radiation. This pigmentation, however, makes them more vulnerable to visual fish predation. Hansson (2004) found that the pigmentation in the copepods decreased when they came in contact with water in which fish had been swimming. Similarly, Tollrian and Heibl (2004) reported reduced pigmentation for *Daphnia*. The pigmentation was the lowest if the UV radiation was low as well. A lower pigmentation increased the mortality due to UV radiation, but decreased the mortality due to fish predation. So, there is a trade-off between risks from predation and ultraviolet radiation.

In contrast to several studies on freshwater plankton, the role of predator infochemicals in marine zooplankton has been investigated by only a few workers (e.g., Strand and Hamner 1990; Bollens et al. 1994; Cieri and Stearns 1999; Hamren and Hansson 1999). There is some evidence that marine copepods respond to mechanical or visual stimuli, rather than to chemicals exuded by predatory fish (Bollens et al. 1994). Several marine plankton species use bioluminescence for communication. However, Cieri and Stearns (1999) and Hamren and Hansson (1999) have demonstrated that marine planktonic crustaceans (copepods, mysidaceans) reduce feeding activity in the presence of fish infochemicals. The resulting reduction in gut fullness is perhaps adaptive in reducing visibility to predators (Bollens and Stearns 1992); this has also been shown for other planktonic organisms (Giguère and Northcote 1987). Thus, predator infochemicals may also play an important role for anti-predator defences in marine plankton.

BENEFITS AND COSTS OF INDUCED DEFENCES

Induced defences provide protection against different predators and allow organisms to adapt phenotypically to multi-predator regimes (Tollrian and Dodson 1999). The defence can be attuned depending on the predator present. Despite the fact that some machinery is required to initiate defence, the costs of inducible defence can be low, because the defence is only initiated in the presence of the predator. Generally, defences are believed to bring about costs that are averted if the predators are absent; otherwise constitutive defences would have been favoured by natural selection (Tollrian and Harvell 1999). For *Daphnia*, researchers were confronted with great difficulties to demonstrate the expected physiological costs of neckteeth formation (Tollrian and Dodson 1999). The costs are reported by some authors to involve trade-offs of life-history reactions to *Chaoborus* infochemicals, i.e., no direct costs result from neckteeth formation for *Daphnia* (Repka and Pihlajamaa 1996). In contrast, morphological defences in ciliates have been found to lead to metabolic costs. In *Euplotes*, protein synthesis is necessary for predator-induced changes in morphology (Kusch and Kuhlmann 1994). These metabolic costs cause increased generation times (Kuhlmann 1992). Consequently, the reduced population growth rates as well as reduced anti-predator morphological changes bring about demographic costs (Kusch and Kuhlmann 1994). Similar demographic costs of anti-predator morphologies are reported for rotifers: at high food concentrations, the predator-induced morph of *Keratella testudo* has less than half of the intrinsic rate of population increase of the non-induced morph (Stemberger 1988). Furthermore, the extent of morphological defence has been observed to correlate with the prevailing predation risk in ciliates (Kusch 1995). The observed adjustment of morphological changes to the actual predation risk indicates that the costs involved are saved when predation risk is reduced or absent. In planktonic algae, colony formation appeared to have direct photosynthetic costs (Verschoor 2005). Furthermore, ecological costs consist of enhanced sinking of colonies out of the euphotic zone (Lürling and Van

Donk 2000). Such environmental costs, i.e., interactions of inducible defences with the environment, might also exist for neckteeth in *Daphnia* (Tollrian and Dodson 1999) and for vertical migration behaviour (Loose and Dawidowicz 1994) and for horizontal migration (Burks et al. 2001). Tollrian and Dodson (1999) state that costs may have been calculated in a simplified context. Phenotypic changes and inducible defences might impose various other costs and limits than simply metabolic costs (DeWitt et al. 1998). One limit may be that adaptations to one predator regime might be unfavourable in the presence of another predator. For example, morphological changes generally increase visibility that might be disadvantageous in the presence of predators that stalk their prey by vision (Tollrian 1995). Furthermore, selection should favour costs that are lower than the benefits and are as low as possible. Thus, costs might be absent or small and difficult to measure but still relevant for prey populations (Tollrian 1995; Tollrian and Dodson 1999).

CASCADING EFFECTS: POPULATION DYNAMICS AND ECOSYSTEM FUNCTIONING

All changes in prey morphology and behaviour in plankton in response to infochemicals from a potential predator will increase the probability of survival for an individual prey organism, but it may also have population- and even system-wide consequences. Diel vertical or horizontal migration of zooplankton in response to infochemicals from planktivorous fish will affect the resource availability for the fish as well as grazing pressure on phytoplankton. Although inducible defences have been investigated extensively at the level of individuals and populations, their importance for population dynamics and ecosystem functioning has hardly been investigated.

Vos et al. (2002) used a combination of individual-based modelling and experimental data from the field and laboratory to show that induced defences in *Daphnia* significantly reduced predation by juvenile perch on *Daphnia* populations during early summer. Induced defences thus prevent overexploitation of the *Daphnia* population by fish and allow the zooplankters to persist. Behavioural defences through diel vertical migration were shown to have a much stronger quantitative effect than defences through changes in life history (Vos et al. 2002). Vos et al. (2004a) predicted that nutrient enrichment could destabilize aquatic food chains when defences in prey are fixed or absent, while such destabilization, the so-called paradox of enrichment, could be absent when prey have inducible defences (Vos et al. 2004a).

Verschoor et al. (2004b) empirically tested the predictions by Vos et al. (2004a), using food chains consisting of inducible defended and undefended algae, herbivorous rotifers and carnivorous rotifers. In enriched food chains with undefended algae, they observed large-amplitude oscillations over several orders of magnitude, which incidentally resulted in extinction of the top predator. On the other hand, food chains with inducible defended algae stabilized immediately after the initial transient phase (Figure 2). Thus, induced defences prevented strong fluctuations and extinctions of higher trophic levels.

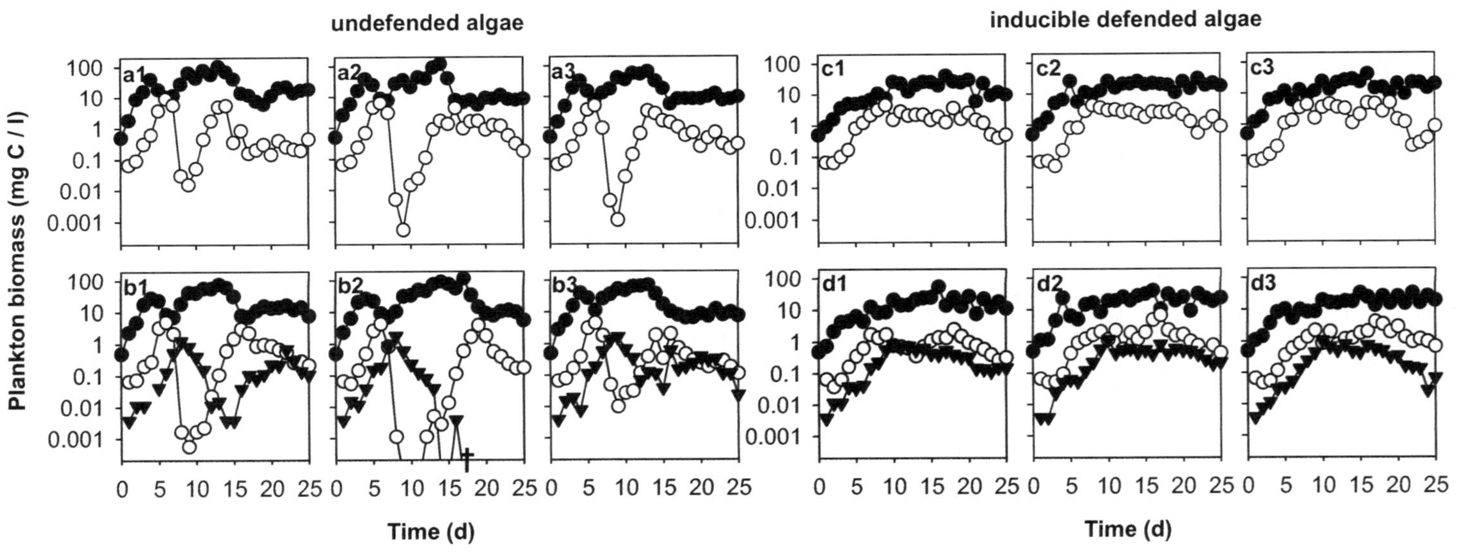

Figure 2. Population dynamics of planktonic food chains in high-phosphorus treatments, with densities expressed as mg Cl^{-1}. Solid circles represent phytoplankton biomass, open circles represent herbivorous zooplankton (Brachionus) and triangles represent carnivorous zooplankton (Asplanchna). ***a, b***. food chains with undefended phytoplankton (Desmodesmus); ***c, d***. food chains with inducible defences in phytoplankton (Scenedesmus); numbers indicate different replicates. Zooplankton extinctions are marked by † (From Verschoor et al. 2004b)

Experiments with more species-rich food webs are needed to get a deeper insight into the defence responses at different trophic levels. The effects of chemical information transfer and induced defences at ecosystem level are still not understood, but induced defences have been predicted to cause all trophic levels to increase under enrichment, a pattern that is consistent with both field and laboratory observations (Vos et al. 2004b). Chemical interactions between planktonic organisms may hamper lake restoration by food-web manipulation but to test this, field studies are necessary.

THE CHEMICAL NATURE AND TRANSPORTATION OF INFOCHEMICALS

Although the presence of infochemicals has been confirmed in many systems, the chemical structures of many of these compounds are still quite obscure. Especially for freshwater environments the chemistry of these compounds is little known. Nonetheless, three kairomones, produced by *Lembadion* (predatory ciliate), *Amoeba* and the flatworm *Stenostomum,* which affect a freshwater ciliate, *Euplotes*, have been described (Kusch 1999; Kusch and Heckmann 1992). These compounds are complex proteins varying in molecular weight from 4.5 kDa to 31.5 kDa.

Most research on infochemicals excreted by predators of *Daphnia* is at present aimed at their characterization. Boriss et al. (1999) reported that trimethylamine (TMA) produced by fish was the compound responsible for the defence response of *Daphnia.* However, in a follow-up study Pohnert and Von Elert (2000) found that *Daphnia* responded to TMA only at unrealistically high concentrations. Furthermore, daphnids continued to exhibit DVM even when TMA was removed from the fish water (Pohnert and Von Elert 2000). Ringelberg and Van Gool (1998) suggested that it is not the fish themselves but bacteria associated with them that produce the infochemical, which triggers the daphnids to migrate. When the fish were treated with antibiotics in solution, so as to immobilize the fish-associated bacteria, the fish water induced significantly less DVM in the daphnids than in controls (fish water untreated with antibiotics). Nonetheless, because some significant biological activity still remained, apparently bacteria cannot be the sole causal factor for the DVM.

The response of *Daphnia* to the *Notonecta* infochemical resembled that of the fish factor (Dodson 1989; Riessen 1999). So far, the *Notonecta* cue has not been characterized. The response to the infochemical from *Chaoborus* fundamentally differed from the cues from fish and *Notonecta* (Dodson 1989; Loose et al. 1993; Riessen 1999). Thus, the chemical compounds involved may be different (Larsson and Dodson 1993; Loose et al. 1993; Riessen 1999). The difference is also clear from the different chemical characteristics (Tollrian and Von Elert 1994).

Until now, the infochemical released by *Daphnia* that causes *Scenedesmus* to form colonies has been poorly described. Wiltshire and Lampert (1999) reported that urea is the infochemical that induces *Scenedesmus* colony formation. However, Lürling and Von Elert (2001) found evidence that contradicts this. Earlier, in 1994, Lampert and co-workers had already obtained negative results with several concentrations of urea (Lampert et al. 1994). Therefore, the *Daphnia* factor is most likely not urea. This is also supported by the fact that urea would be formed by the general metabolism, while Von Elert and Franck (1999) presented evidence that the infochemical originates from a particular metabolic reaction. Kaler et al. (2000) propose oligonucleotides or nucleic acids and possibly peptides as possible structures for the consumer-released factor, solely based on UV spectra. This view is, however, not supported by the experimental work of Lampert et al. (1994) and Von Elert and Franck (1999). The chemical cue can be enriched on SPE cartridges (C_{18}) (Lampert et al. 1994; Von Elert and Franck 1999) and is heat- and pH-stable, water-soluble and non-volatile, and has a molecular weight <0.5 kDa (Lampert et al. 1994). Furthermore, its retention by a strongly basic anion exchanger implies that it is an anionic compound and thus excludes urea (Von Elert and Franck 1999). Other experiments hint at the presence of hydrophilic groups, possibly a carboxyl group, and olefinic characteristics. The extract subjected to HPLC showed activity in only one fraction (Van Holthoon 2004). Because the extracts of mechanically crushed *Daphnia* or *Scenedesmus* did not show colonizing activity in the latter (Lampert et al. 1994; Von Elert and Franck 1999) an interaction between *Scenedesmus* and *Daphnia* may be needed to initiate production of the active compound (Lürling and Van Donk 1996). Lürling (1999) attributed colony formation to feeding activity of *Daphnia* rather than merely to the presence of daphnids, as starved animals did not induce colony formation in *Scenedesmus* (Lürling and Van Donk 1996). In analogy to the fish factor (Ringelberg and Van Gool 1998), bacteria in the gut of *Daphnia* may release such an infochemical (Fink 2001).

Once the chemicals responsible for transferring information between organisms are identified, important research tasks will be to assess how these chemicals are transported in the water and how important turbulence is for their dispersion. Until now, most studies have been performed in standing water or at laminar flow in laboratory settings, whereas in natural situations turbulent mixing dilutes chemical stimuli and creates patchiness in odour distributions resulting in a much more complex olfactory landscape (e.g., Zimmer-Faust et al. 1995; Zimmer et al. 1999), much as is the case with airborne infochemicals.

ALLELOPATHIC INTERACTIONS

Aquatic macrophytes have long been suspected of suppressing phytoplankton growth through the excretion of growth-inhibiting chemical substances (Van Donk and Van de Bund 2002). The production and excretion of chemicals by aquatic macrophytes could be an effective defence strategy against other photosynthetic organisms, epiphyton and phytoplankton, which compete with macrophytes for light and nutrients. This is, however, not an induced defence and, in contrast with the nature of the infochemicals inducing defence, more knowledge is available about the structure of allelopathic chemicals. These substances belong to rather different chemical classes such as sulphur compounds, polyacetylenes, polyphenols and oxygenated fatty acids (Gross 1999). Submerged macrophytes such as *Ceratophyllum, Stratiotes, Chara* and *Myriophyllum* may strongly inhibit algal growth, and sensitivity of different algal species to these chemicals differs (Körner and Nicklisch 2002; Mulderij et al. 2003). Consequently, these allelopathic substances may change the composition and dynamics of the phytoplankton community.

Allelopathic interactions have also been reported between phytoplankton species. A review about allelopathy in phytoplankton and the biochemical, ecological and evolutionary aspects, is given by Legrand et al. (2003). They state that chemical interactions, specifically allelopathy, are an important part of phytoplankton competition. For example, *Microcystis* is able to delay the start of the bloom of another phytoplankter *Peridinium* by release of a chemical compound, which also inhibits the growth of other cyanobacteria like *Nostoc* and *Anabaena* directly (Singh et al. 2001; Sukenik et al. 2002).

ATTRACTION TO FOOD BY MEANS OF INFOCHEMICALS

Chemical cues from potential food algae could be important for their consumers because they provide information on the food quality (Larsson and Dodson 1993). So, according to Van Gool and Ringelberg (1996), algae-associated odours could be detected by *Daphnia*. The daphnids were observed to swim towards the edible green alga *Scenedesmus*, but not towards toxic, less edible cyanobacteria. Haney et al. (1994) demonstrated that the food intake of the daphnids was reduced on exposing them to chemicals released by cyanobacteria. In fact, one would expect *Daphnia* to avoid areas where potentially harmful cyanobacteria are present. In contrast, in food-gradient experiments *Daphnia* strongly aggregated in zones with intermediate food levels but avoided zones with high food levels (Neary et al. 1994). The mechanism used by *Daphnia* to locate these regions is probably related to the concentration of algal cells rather than the presence of algal odours. So, several factors, including olfaction, affect *Daphnia*–algal interactions (Roozen and Lürling 2001).

A chemical attraction has also been found between a parasitic fungus and its algal host, a diatom, *Asterionela formosa* (Van Donk 1989; Ibelings et al. 2004). This fungal infection is very host-specific. The spores swim towards their algal host, attach themselves and penetrate the host cells. These spores form sporangia that mature and produce spores again. The zoospores can only locate a new host algal cell in the light, i.e., when the alga produces an exudate during photosynthesis. In the dark such exudates are not produced (Canter and Jaworski 1981).

MULTITROPHIC INDIRECT DEFENCE

Chemical attraction can also play a role in multitrophic indirect defences. For terrestrial ecosystems it has been found that in addition to a direct defence in plants, like a herbivore-induced toxin in the leaves, there is an indirect multitrophic defence. In response to grazing by caterpillars, for example, plants produce volatiles that attract the parasitic wasps that parasitize the caterpillars (Dicke 1999). Furthermore, some plants can induce their neighbouring plants to produce volatiles that attract predators, causing increase in predation on the herbivores (Dicke and Bruin 2001).

Indirect multitrophic defences have also been described for marine ecosystems. Dimethylsulfide is produced by the alga *Phaeocystis* when micro-zooplankton is grazing on them (Strom and Wolfe 2001). As a reaction copepods are attracted towards the microzooplankton and consume it, thereby reducing their grazing pressure on *Phaeocystis* and thus indirectly protecting the alga. Dimethylsulfide, which is volatile and will reach the surface of the water, will attract fish-eating birds to high-production areas with a high concentration of fish predating on the copepods (Steinke et al. 2002). This phenomenon will indirectly also result in protection of the alga.

The question that arises is: are such indirect, induced defence mechanisms present in freshwater ecosystems? Kagami et al. (2004) found that *Daphnia* can eat fungal zoospores and protect the alga *Asterionella* from fungal infection. These spores are more readily eaten when they are concentrated around their host and conceivably the algae can exude some infochemical attracting *Daphnia* to defend themselves. A similar explanation can be proposed for the interaction between cyanobacteria, their parasitic virus and the flagellates eating on this virus. Murray (1995) found that cyanobacteria excrete organic substances that attract these virus-eating flagellates. Further research is needed to test these hypotheses about multitrophic indirect defences in freshwaters.

CONCLUDING REMARKS

From the preceding sections it is clear that planktonic interactions are highly variable and complex. We are now more aware of the infochemicals that are involved in planktonic interactions and have begun to accumulate knowledge about the nature of these infochemicals. Nonetheless, we are still not able to identify and isolate them. Apparently, chemical communication mediates both predation and interspecific competition, and we know that induced defences moderate strong population fluctuations and the local extinctions of consumers that may result from these. However, we do not know yet what and how much impact chemical communication has on ecosystem functioning. One may hypothesize that chemical interactions between planktonic organisms hamper lake restoration by food-web manipulation, but field studies are necessary to test this.

Finally, I believe that future research in chemical ecology will help pave the way to a better understanding of species composition, top-down control of algae and the structure of aquatic food webs.

ACKNOWLEDGEMENTS

This paper was improved by comments from M. Dicke, R. Tollrian and W. Takken.

REFERENCES

Benndorf, J. and Henning, M., 1989. *Daphnia* and toxic blooms of *Microcystis aeruginosa* in Bautzen Reservoir (GDR). *Internationale Revue der Gesamten Hydrobiologie,* 74 (3), 233-248.

Bollens, S.M., Frost, B.W. and Cordell, J.R., 1994. Chemical, mechanical and visual cues in the vertical migration behavior of the marine planktonic copepod *Acartia hudsonica. Journal of Plankton Research,* 16 (5), 555-564.

Bollens, S.M. and Stearns, D.E., 1992. Predator-induced changes in the diel feeding cycle of a planktonic copepod. *Journal of Experimental Marine Biology and Ecology,* 156 (2), 179-186.

Boriss, H., Boersma, M. and Wiltshire, K.H., 1999. Trimethylamine induces migration of waterfleas. *Nature,* 398 (6726), 382.

Brönmark, C. and Hansson, L.A., 2000. Chemical communication in aquatic systems: an introduction. *Oikos,* 88 (1), 103-109.

Burks, R.L., Jeppesen, E. and Lodge, D.M., 2000. Macrophyte and fish chemicals suppress *Daphnia* growth and alter life-history traits. *Oikos,* 88 (1), 139-147.

Burks, R.L., Jeppesen, E. and Lodge, D.M., 2001. Pelagic prey and benthic predators: impact of odonate predation on *Daphnia. Journal of the North American Benthological Society,* 20 (4), 615-628.

Burns, C.W., 1968. The relationship between body size of filter feeding Cladocera and the maximum size of particle ingested. *Limnology and Oceanography,* 13 (4), 675-678.

Canter, H.M. and Jaworski, G.H.M., 1981. The effect of light and darkness upon infection of *Asterionella formosa* Hassal by the chytrid *Rhizophydium planktonicum* Canter emend. *Annals of Botany,* 47, 13-30.

Cieri, M.D. and Stearns, D.E., 1999. Reduction of grazing activity of two estuarine copepods in response to the exudate of a visual predator. *Marine Ecology Progress Series,* 177, 157-163.

DeMott, W.R., 1995. The influence of prey hardness on *Daphnia*'s selectivity for large prey. *Hydrobiologia,* 307 (1/3), 127-138.

DeWitt, T.J., Sih, A. and Wilson, D.S., 1998. Costs and limits of phenotypic plasticity. *Trends in Ecology and Evolution,* 13 (2), 77-81.

Dicke, M., 1999. Evolution of induced indirect defence of plants. *In:* Tollrian, R. and Harvell, C.D. eds. *The ecology and evolution of inducible defenses*. Princeton University Press, Princeton, 62-88.

Dicke, M. and Bruin, J., 2001. Chemical information transfer between plants: back to the future. *Biochemical Systematics and Ecology,* 29 (10), 981-994.

Dicke, M. and Sabelis, M.W., 1988. Infochemical terminology: based on cost-benefit analysis rather than origin of compounds? *Functional Ecology,* 2 (2), 131-139.

Dicke, M. and Sabelis, M.W., 1992. Costs and benefits of chemical information conveyance: proximate and ultimate factors. *In:* Roitberg, B.D. and Isman, M.B. eds. *Insect chemical ecology: an evolutionary approach*. Chapman & Hall, New York, 122-155.

Dodson, S.I., 1988. The ecological role of chemical stimuli for the zooplankton predator-avoidance behavior in *Daphnia. Limnology and Oceanography,* 33 (6, part 2), 1431-1439.

Dodson, S.I., 1989. The ecological role of chemical stimuli for the zooplankton predator-induced morphology in *Daphnia. Oecologia,* 78 (3), 361-367.

Fink, P., 2001. *Untersuchungen zur Coenobieninduktion in Scenedesmus durch Daphnia.* MSc dissertation, Universität Konstanz.

Giguère, L.A. and Northcote, T.G., 1987. Ingested prey increase risks of visual predation in transparent *Chaoborus* larvae. *Oecologia,* 73 (1), 48-52.

Gilbert, J.J., 1966. Rotifer ecology and embryological induction. *Science,* 151 (715), 1234-1237.

Grant, J.W.G. and Bayly, I.A.E., 1981. Predator induction of crests in morphs of the *Daphnia carinata* complex. *Limnology and Oceanography,* 26 (2), 201-218.

Gross, E.M., 1999. Allelopathy in benthic and littoral areas: case studies on allelochemicals from benthic cyanobacteria and submerged macrophytes. *In:* Inderjit, Dakshini, K.M.M. and Chester, L.F. eds. *Principles and practices in plant ecology: allelochemical interactions*. CRC Press, Boca Raton, 179-199.

Gulati, R.D., Bronkhorst, M. and Van Donk, E., 2001. Feeding in *Daphnia galeata* on *Oscillatoria limnetica* and on detritus derived from it. *Journal of Plankton Research,* 23 (7), 705-718.

Hamren, U. and Hansson, S., 1999. A mysid shrimp (*Mysis mixta*) is able to detect the odour of its predator (*Clupea harengus*). *Ophelia,* 51 (3), 187-191.
Haney, J.F., Forsyth, D.J. and James, M.R., 1994. Inhibition of zooplankton filtering rates by dissolved inhibitors produced by naturally occurring cyanobacteria. *Archiv für Hydrobiologie,* 132 (1), 1-13.
Hansson, L.A., 1996. Behavioural response in plants: adjustment in algal recruitment induced by herbivores. *Proceedings of the Royal Society of London. Series B. Biological Sciences,* 263 (1374), 1241-1244.
Hansson, L.A., 2000. Synergistic effects of food chain dynamics and induced behavioral responses in aquatic ecosystems. *Ecology,* 81 (3), 842-851.
Hansson, L.A., 2004. Plasticity in pigmentation induced by conflicting threats from predation and UV radiation. *Ecology,* 85 (4).
Hessen, D.O. and Van Donk, E., 1993. Morphological changes in *Scenedesmus* induced by substances released from *Daphnia*. *Archiv für Hydrobiologie,* 127 (2), 129-140.
Hutchinson, G.E., 1967. *A treatise on Limnology. Vol. II. Introduction to lake biology and the limnoplankton*. John Wiley and Sons, New York.
Ibelings, B.W., De Bruin, A., Kagami, M., et al. 2004. Host parasite interactions between freshwater phytoplankton and chytrid fungi (Chytridiomycota). *Journal of Phycology*.
Jang, M-H., Ha, K., Joo, G-J., et al. 2003. Toxin production of cyanobacteria is increased by exposure to zooplankton. *Freshwater Biology,* 48 (9), 1540-1550.
Kagami, M., Van Donk, E., De Bruin, A., et al. 2004. *Daphnia* can protect diatoms from fungal parasitism. *Limnology and Oceanography,* 49 (3), 680-685.
Kaler, V.L., Bulko, O.P., Reshetnikov, V.N., et al. 2000. Changes in the morphostructure of *Scenedesmus acutus* and culture growth rate induced by the exudate of primary consumer *Daphnia magna*. *Russian Journal of Plant Physiology,* 47 (5), 698-705.
Körner, S. and Nicklisch, A., 2002. Allelopathic growth inhibition of selected phytoplankton species by submerged macrophytes. *Journal of Phycology,* 38 (5), 862-871.
Krueger, D.A. and Dodson, S.I., 1981. Embryological induction and predation ecology in *Daphnia pulex*. *Limnology and Oceanography,* 26 (2), 219-223.
Kuhlmann, H.W., 1992. Benefits and costs of predator-induced defences in *Euplotes*. *Journal of Protozoology,* 39, 49A.
Kuhlmann, H.W. and Heckmann, K., 1985. Interspecific morphogens regulating prey-predator relationships in protozoa. *Science,* 227 (4692), 1347-1349.
Kusch, J., 1995. Adaptation of inducible defense in *Euplotes daidaleos* (Ciliophora) to predation risks by various predators. *Microbial Ecology,* 30 (1), 79-88.
Kusch, J., 1999. Self-recognition as the original function of an amoeban defense-inducing kairomone. *Ecology,* 80 (2), 715-720.
Kusch, J. and Heckmann, K., 1992. Isolation of the *Lembadion*-factor, a morphogenetically active signal that induces *Euplotes* cells to change from their ovoid form into a larger lateral winged morph. *Developmental Genetics,* 13 (3), 241-246.
Kusch, J. and Kuhlmann, H.W., 1994. Cost of *Stenostomum*-induced morphological defence in the ciliate *Euplotes octocarinatus*. *Archiv für Hydrobiologie,* 130 (3), 257-267.
Lampert, W., 1987. Feeding and nutrition in *Daphnia*. *Memorie dell'Istituto Italiano di Idrobiologia,* 45, 143-192.
Lampert, W., Rothhaupt, K.O. and Von Elert, E., 1994. Chemical induction of colony formation in a green alga (*Scenedesmus acutus*) by grazers (*Daphnia*). *Limnology and Oceanography,* 39 (7), 1543-1550.
Larsson, P. and Dodson, S.I., 1993. Chemical communication in planktonic animals. *Archiv für Hydrobiologie,* 129 (2), 129-155.
Lass, S. and Spaak, P., 2003. Chemically induced anti-predator defences in plankton: a review. *Hydrobiologia,* 491, 221-239.
Legrand, C., Rengefors, K., Fistarol, G.O., et al. 2003. Allelopathy in phytoplankton: biochemical, ecological and evolutionary aspects. *Phycologia,* 42 (4), 406-419.
Loose, C.J. and Dawidowicz, P., 1994. Trade-offs in diel vertical migration by zooplankton: the costs of predator avoidance. *Ecology,* 75 (8), 2255-2263.
Loose, C.J., Von Elert, E. and Dawidowicz, P., 1993. Chemically-induced diel vertical migration in *Daphnia*: a new bioassay for kairomones exuded by fish. *Archiv für Hydrobiologie,* 126 (3), 329-337.

Lürling, M., 1999. *The smell of water: grazer-induced colony formation in Scenedesmus*. PhD Thesis, Wageningen University, Wageningen.
Lürling, M. and Van Donk, E., 1996. Zooplankton-induced unicell-colony transformation in *Scenedesmus acutus* and its effect on growth of herbivore *Daphnia*. *Oecologia,* 108 (3), 432-437.
Lürling, M. and Van Donk, E., 2000. Grazer-induced colony formation in *Scenedesmus*: are there costs to being colonial? *Oikos,* 88 (1), 111-118.
Lürling, M. and Von Elert, E., 2001. Colony formation in *Scenedesmus*: no contribution of urea in induction by a lipophilic *Daphnia* exudate. *Limnology and Oceanography,* 46 (7), 1809-1813.
Lynch, M., 1980. *Aphanizomenon* blooms: alternate control and cultivation by *Daphnia pulex*. *In:* Kerfoot, W.C. ed. *Evolution and ecology of zooplankton communities*. University Press of New England, Hanover, 299-304.
Malej, A. and Harris, R.P., 1993. Inhibition of copepod grazing by diatom exudates: a factor in the development of mucus aggregates? *Marine Ecology Progress Series,* 96 (1), 33-42.
Mulderij, G., Van Donk, E. and Roelofs, J.G.M., 2003. Differential sensitivity of green algae to allelopathic substances from *Chara*. *Hydrobiologia,* 491, 261-271.
Murray, A., 1995. Phytoplankton exudation: exploitation of the microbial loop as a defence against algal viruses. *Journal of Plankton Research,* 17 (5), 1079-1094.
Neary, J., Cash, K. and McCauley, E., 1994. Behavioural aggregation of *Daphnia pulex* in response to food gradients. *Functional Ecology,* 8 (3), 377-383.
Pijanowska, J. and Kowalczewski, A., 1997. Predators can induce swarming behaviour and locomotory responses in *Daphnia*. *Freshwater Biology,* 37 (3), 649-656.
Pohnert, G. and Von Elert, E., 2000. No ecological relevance of trimethylamine in fish-*Daphnia* interactions. *Limnology and Oceanography,* 45 (5), 1153-1156.
Porter, K.G., 1975. Viable gut passage of gelatinous green algae ingested by *Daphnia*. *Verhandlungen der Internationale Vereinigung fur Theoretische und Angewandte Limnologie,* 19, 2840-2850.
Repka, S. and Pihlajamaa, K., 1996. Predator-induced phenotypic plasticity in *Daphnia pulex*: uncoupling morphological defenses and life history shifts. *Hydrobiologia,* 339 (1/3), 67-71.
Riessen, H.P., 1999. Predator-induced life history shifts in *Daphnia*: a synthesis of studies using meta-analysis. *Canadian Journal of Fisheries and Aquatic Sciences,* 56 (12), 2487-2494.
Ringelberg, J. and Van Gool, E., 1998. Do bacteria, not fish, produce 'fish kairomone'? *Journal of Plankton Research,* 20 (9), 1847-1852.
Roozen, F. and Lürling, M., 2001. Behavioural response of *Daphnia* to olfactory cues from food, competitors and predators. *Journal of Plankton Research,* 23 (8), 797-808.
Singh, D.P., Tyagi, M.B., Kumar, A., et al. 2001. Antialgal activity of a hepatotoxin-producing cyanobacterium *Microcystis aeruginosa*. *World Journal of Microbiology and Biotechnology,* 17 (1), 15-22.
Slusarczyk, M., 1999. Predator-induced diapause in *Daphnia magna* may require two chemical cues. *Oecologia,* 119 (2), 159 165.
Steinke, M., Malin, G. and Liss, P.S., 2002. Trophic interactions in the sea: an ecological role for climate relevant volatiles? *Journal of Phycology,* 38 (4), 630-638.
Stemberger, R.S., 1988. Reproductive costs and hydrodynamic benefits of chemically induced defenses in *Keratella testudo*. *Limnology and Oceanography,* 33 (4, part 1), 593-606.
Sterner, R.W. and Smith, R.F., 1993. Clearance, ingestion and release of N and P by *Daphnia obtusa* feeding on *Scenedesmus acutus* of varying quality. *Bulletin of Marine Science,* 53 (1), 228-239.
Strand, S.W. and Hamner, W.M., 1990. Schooling behavior of antarctic krill (*Euphausia superba*) in laboratory aquaria: reactions to chemical and visual stimuli. *Marine Biology,* 106 (3), 355-360.
Strom, S.L. and Wolfe, G.V., 2001. Phytoplankton chemical signals influence herbivory by protist grazers. *Journal of Phycology,* 37 (3), 47-48.
Stutzman, P., 1995. Food quality of gelatinous colonial chlorophytes to the freshwater zooplankters *Daphnia pulicaria* and *Diaptomus oregonensis*. *Freshwater Biology,* 34 (1), 149-153.
Sukenik, A., Eshkol, R., Livne, A., et al. 2002. Inhibition of growth and photosynthesis of the dinoflagellate *Peridinium gatunense* by *Microcystis* sp. (cyanobacteria): a novel allelopathic mechanism. *Limnology and Oceanography,* 47 (6), 1656-1663.
Tollrian, R., 1994. Fish-kairomone induced morphological changes in *Daphnia lumholtzi* (Sars.). *Archiv für Hydrobiologie,* 130 (1), 69-75.
Tollrian, R., 1995. Predator-induced morphological defenses: costs, life history shifts, and maternal effects in *Daphnia pulex*. *Ecology,* 76 (6), 1691-1705.

Tollrian, R. and Dodson, S.I., 1999. Inducible defenses in Cladocera: constraints, costs, and multipredator environments. *In:* Tollrian, R. and Harvell, C.D. eds. *The ecology and evolution of inducible defenses*. Princeton University Press, Princeton, 177-202.
Tollrian, R. and Harvell, C.D. (eds.), 1999. *The ecology and evolution of inducible defenses*. Princeton University Press, Princeton.
Tollrian, R. and Heibl, C., 2004. Phenotypic plasticity in pigmentation in *Daphnia* induced by UV radiation and fish kairomones. *Functional Ecology,* 18 (4), 497-502.
Tollrian, R. and Von Elert, E., 1994. Enrichment and purification of Chaoborus kairomone from water: further steps toward its chemical characterization. *Limnology and Oceanography,* 39 (4), 788-796.
Trainor, F.R. and Egan, P.F., 1991. Discovering the various ecomorphs of *Scenedesmus*: the end of a taxonomic era. *Archiv für Protistenkunde,* 139 (1/4), 125-132.
Van Donk, E., 1989. The role of parasitism of phytoplankton succession: a review. *In:* Sommer, U. ed. *Plankton ecology: succession in plankton communities*. Springer Verlag, Berlin, 171-194.
Van Donk, E. and Hessen, D.O., 1993. Grazing resistance in nutrient-stressed phytoplankton. *Oecologia,* 93 (4), 508-511.
Van Donk, E., Lürling, M., Hessen, D.O., et al. 1997. Altered cell wall morphology in nutrient-deficient phytoplankton and its impact on grazers. *Limnology and Oceanography,* 42 (2), 357-364.
Van Donk, E., Lürling, M. and Lampert, W., 1999. Consumer-induced changes in phytoplankton: inducibility, costs, benefits and the impact on grazers. *In:* Tollrian, R. and Harvell, C.D. eds. *Ecology and evolution of inducible defenses*. Princeton University Press, Princeton, 89-103.
Van Donk, E. and Van de Bund, W.J., 2002. Impact of submerged macrophytes including charophytes on phyto- and zooplankton communities: allelopathy versus other mechanisms. *Aquatic Botany,* 72 (3/4), 261-274.
Van Gool, E. and Ringelberg, J., 1996. Daphnids respond to algae-associated odours. *Journal of Plankton Research,* 18 (1), 197-202.
Van Holthoon, F.L., 2004. *Isolation and identification of kairomone(s) in the Daphnia/Scenedesmus system*. PhD dissertation, Wageningen University, Wageningen.
Verschoor, A.M., 2005. *Hard to handle: induced defences in plankton*. PhD dissertation, Radboud University Nijmegen, Nijmegen. [http://webdoc.ubn.ru.nl/mono/v/verschoor_a/hardtoha.pdf].
Verschoor, A.M., Van der Stap, I., Helmsing, N.R., et al. 2004a. Inducible colony formation within the Scenedesmaceae: adaptive responses to infochemicals from two different herbivore taxa. *Journal of Phycology,* 40 (5), 808-814.
Verschoor, A.M., Vos, M. and Van der Stap, I., 2004b. Inducible defences prevent strong population fluctuations in bi- and tritrophic food chains. *Ecology Letters,* 7 (12), 1143-1148.
Von Elert, E. and Franck, A., 1999. Colony formation in *Scenedesmus*: grazer-mediated release and chemical features of the infochemical. *Journal of Plankton Research,* 21 (4), 789-804.
Vos, M., Flik, B.J.G., Vijverberg, J., et al. 2002. From inducible defences to population dynamics: modelling refuge use and life history changes in *Daphnia*. *Oikos,* 99 (2), 386-396.
Vos, M., Kooi, B.W., DeAngelis, D.L., et al. 2004a. Inducible defences and the paradox of enrichment. *Oikos,* 105 (3), 471-480.
Vos, M., Verschoor, A.M., Kooi, B.W., et al. 2004b. Inducible defenses and trophic structure. *Ecology,* 85 (10), 2783-2794.
Wiltshire, K., Boersma, M. and Meyer, B., 2003. Grazer-induced changes in the desmid *Staurastrum*. *Hydrobiologia,* 491, 255-260.
Wiltshire, K.H. and Lampert, W., 1999. Urea excretion by *Daphnia*: A colony-inducing factor in *Scenedesmus*? *Limnology and Oceanography,* 44 (8), 1894-1903.
Wolfe, G.V., Steinke, M. and Kirst, G.O., 1997. Grazing-activated chemical defence in a unicellular marine alga. *Nature,* 387 (6636), 894-897.
Wood, A.M. and Leatham, T., 1992. The species concept in phytoplankton ecology. *Journal of Phycology,* 28, 723-729.
Yoshida, T., Jones, L.E., Ellner, S.P., et al. 2003. Rapid evolution drives ecological dynamics in a predator-prey system. *Nature,* 424 (6946), 303-306.
Zimmer-Faust, R.K., Finelli, C.M., Pentcheff, N.D., et al. 1995. Odor plumes and animal navigation in turbulent water flow: a field study. *Biological Bulletin Woods Hole,* 188 (2), 111-116.
Zimmer, R.K., Commins, J.E. and Browne, K.A., 1999. Regulatory effects of environmental chemical signals on search behavior and foraging success. *Ecology,* 80 (4), 1432-1446.

CHAPTER 11

PLANT VOLATILES YIELDING NEW WAYS TO EXPLOIT PLANT DEFENCE

JOHN A. PICKETT[*,#], TOBY J.A. BRUCE[#], KEITH CHAMBERLAIN[#], AHMED HASSANALI[##], ZEYAUR R. KHAN[##], MICHAELA C. MATTHES[#], JOHNATHAN A. NAPIER[#], LESLEY E. SMART[#], LESTER J. WADHAMS[#] AND CHRISTINE M. WOODCOCK[#]

** Corresponding author. E-mail: john.pickett@bbsrc.ac.uk*
[#] Rothamsted Research, Harpenden, Hertfordshire, AL5 2JQ, UK
[##] International Centre of Insect Physiology and Ecology, PO Box 30772, Nairobi, Kenya

Abstract. When plants are damaged, they produce semiochemicals which can act as repellents for herbivorous pests and as attractants for organisms antagonistic to these pests, e.g., predators and parasitic wasps. Plants can also produce signals that warn other plants of impending attack. From this range of phenomena, it is possible to identify new ways to control pests. Although, in the past, we have needed to deploy such approaches by applying slow-release formulations of semiochemicals to crop plants, we can now use the plants themselves as a source of these semiochemicals. This may be achieved by using inducing agents, or a new range of natural product plant activators, to 'switch on' plant defence prior to attack. This paper considers the identification of new plant activators. In addition, practical use of plants releasing semiochemicals to ward off pest attack, to ensnare the attackers, and to attract beneficial insects that will attack the pests, is demonstrated by use of the stimulo-deterrent diversionary ('push-pull') strategy that has been developed for management of stem-borer moths in Africa.
Keywords: semiochemical; push-pull; non-host; electrophysiology; *cis*-jasmone; jasmonate

INTRODUCTION

We now know that attraction of insects to plants and other host organisms involves detection of specific semiochemicals (natural signal chemicals mediating changes in behaviour and development) (Nordlund and Lewis 1976; Dicke and Sabelis 1988), or specific ratios of semiochemicals. We have also learned, more recently, that the avoidance of unsuitable hosts can involve the detection of specific semiochemicals, or mixtures of semiochemicals, associated with non-host taxa (Hardie et al. 1994; Pettersson et al. 1994). During host alternation by many pest aphids, there can be repulsion away from a host that is not suitable for use at that developmental stage.

M. Dicke and W. Takken (eds.), Chemical Ecology: From Gene to Ecosystem, 161-173.

For example, the winter, or primary, hosts of aphids can produce compounds that repel the spring morphs on their migration back to the summer, or secondary, hosts. Similar phenomena can be observed during colonization by a herbivorous insect, because the plant releases signals indicating that it is already infested and is therefore less suitable as a host. These signals can repel other incoming insects, but can also increase foraging by predators and parasitic wasps. The first interaction with the semiochemicals involved in these types of non-host recognition is usually on the insect antenna. Therefore, by using electroantennography (EAG) or single-cell recording (SCR) from individual olfactory neurons, coupled to high-resolution gas chromatography (GC), we can identify the compounds involved (Pickett et al. 1992).

SEMIOCHEMICALS AS REPELLENTS

Using plants upon which herbivores are feeding, and investigating, by GC-EAG or GC-SCR, the volatile compounds released, it is possible to identify a range of compounds that are electrophysiologically active and which may subsequently prove, in behavioural assays, to be repellents for insect pests. These compounds can also be effective in increasing foraging activity by predators and parasitoids that attack the pests. The compounds involved come from a wide range of biosynthetic pathways, but prominent amongst these are the isoprenoid and lipoxygenase pathways. For example, monoterpenes such as (*E*)-ocimene, and sesquiterpenes such as (-)-germacrene D, can be produced by plants and cause repellency to herbivores (Bruce et al. 2005). However, it is difficult to deploy these chemicals in the field as there is no long-lasting effect and the chemicals themselves are highly volatile and unstable. Heterologous expression of the genes associated with biosynthesis of these compounds has been attempted, but it is often very difficult to obtain useful expression rates, or at least expression that leads to useful production of these compounds (Aharoni et al. 2003). Recently, we have found that the heterologous expression of an (*E*)-β-farnesene synthase in *Arabidopsis thaliana* can be accomplished so that large amounts of (*E*)-β-farnesene are produced, which can affect aphids and their parasitoids (Beale et al. in prep.).

STRESS-RELATED SEMIOCHEMICALS

Methyl salicylate has been identified as a stress-related plant semiochemical, and most insects that we have examined, including some haematophagous insects, show strong electrophysiological responses to this compound. The cereal aphids *Rhopalosiphum padi*, *Sitobion avenae* and *Metopolophium dirhodum* have, in an olfactory organ (the primary rhinarium) on the sixth antennal segment, a specific olfactory neuron for methyl salicylate (Pettersson et al. 1994). This compound, as predicted, is associated with avoidance of cereal crops treated with a slow-release formulation of the material. Thus, in spring field trials, methyl salicylate applied to wheat significantly reduced (by 30-40%) the overall number of aphids colonizing the crop (Pettersson et al. 1994). Methyl salicylate is biosynthetically related to salicylic acid, a signal of systemically acquired resistance (Lucas 1999). This may indicate that the plant is upregulating defence pathways associated with hormonal activity of salicylate and could thereby present difficulties for colonization by herbivores. However, in these trials, the effect was short-lived and the formulation needed to continue to release to provide ongoing field activity.

INDUCTION OF PLANT DEFENCE BY METHYLATED PLANT HORMONES

In addition to direct effects on herbivores, methyl salicylate has been shown, when applied aerially to plants, to induce defence against fungal pathogens (Shulaev et al. 1997). However, a great deal of attention has been directed towards the jasmonate pathway (Figure 1), which is part of the lipoxygenase pathway referred to above. Again, jasmonic acid can act internally as a plant hormone associated with a damage/stress response but, when methylated (i.e., methyl jasmonate, Figure 1), can be released by the plant and, whether naturally or not, will certainly have an effect on intact plants by upregulating defence-related and other genes (Farmer and Ryan 1990; Doughty et al. 1995; Karban et al. 2000; Preston et al. 2002). Unfortunately, a large number of genes are influenced and this can have a deleterious effect on plant development and yields for agricultural crops. Although methylation converts plant hormones such as salicylate and jasmonate to volatile compounds with potential for external signalling, there are other possible mechanisms. From the jasmonate pathway, such an alternative was discovered initially by looking at the chemical ecology of host alternation in aphids.

SEEING *CIS*-JASMONE IN A NEW WAY

When we were studying the host alternation semiochemistry of the lettuce aphid, *Nasonovia ribis-nigri*, we found, as predicted from the above hypothesis, that the spring migrants were repelled by their winter hosts (members of the Saxifragiaceae, e.g., the blackcurrant, *Ribes nigrum*) and that these semiochemicals could act as repellents for such migrants searching for the summer host, lettuce, *Lactuca sativa* (Asteraceae). However, the mixture of semiochemicals contained *cis*-jasmone, which is also involved in the jasmonate pathway (Figure 1). It has been suggested that *cis*-jasmone is a metabolic product of jasmonate and represents a sink for this pathway (Koch et al. 1997), but the behavioural response from *N. ribis-nigri* was very pronounced with this compound alone. A specific olfactory neuron was identified which responded exclusively to *cis*-jasmone, with virtually no response from methyl jasmonate at orders of magnitude greater stimulus concentrations, even though *cis*-jasmone and methyl jasmonate have a close structural resemblance (Figure 1) (Birkett et al. 2000). *cis*-Jasmone was also found to be a repellent for the damson-hop aphid *Phorodon humuli*, taxonomically very different in terms of having a *Prunus* species (Rosaceae) as its primary host and, as a secondary host, the hop *Humulus lupulus* (Cannabiaceae). It was also found that *cis*-jasmone increased attraction and searching behaviour by an aphid predator, the seven-spot ladybird, *Coccinella septempunctata*.

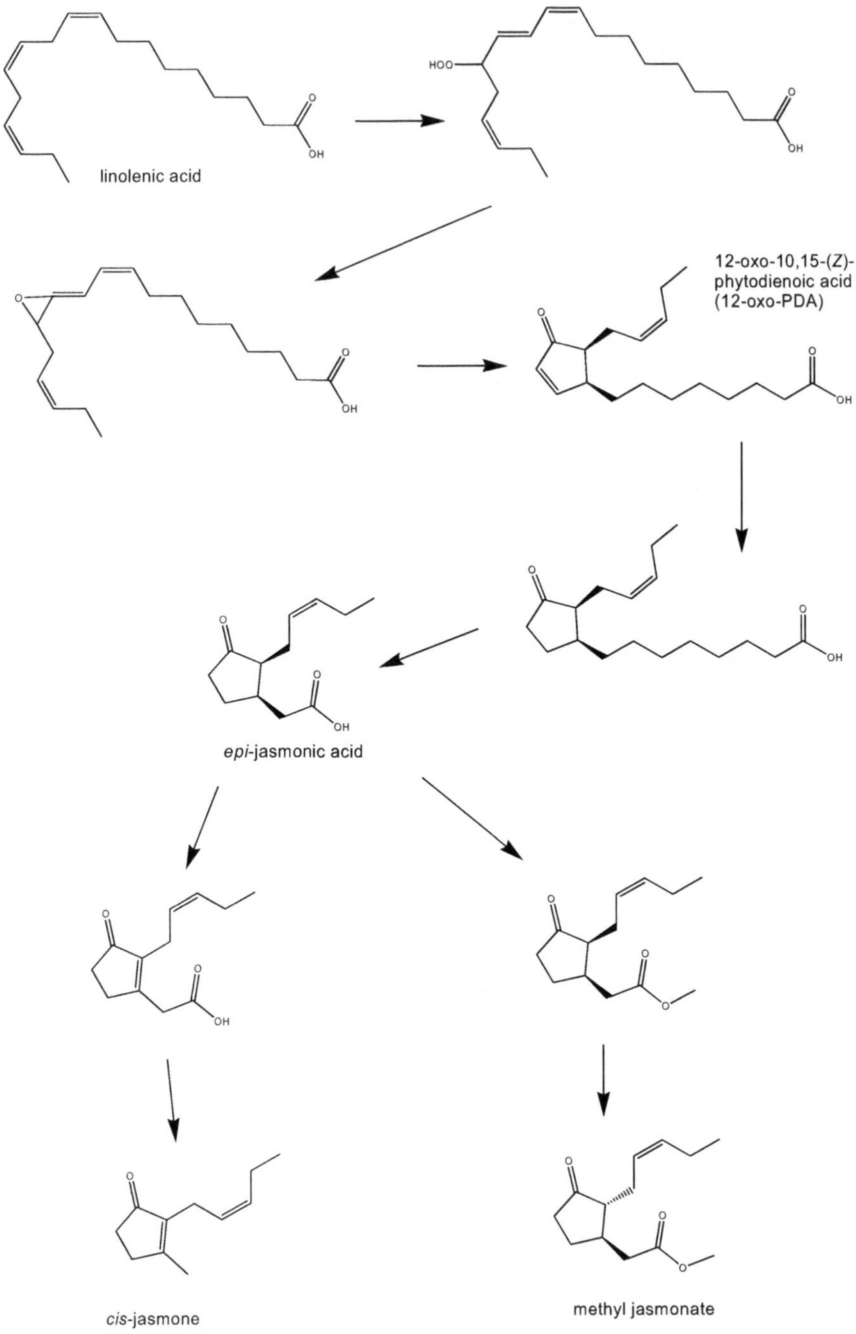

Figure 1. *Biosynthesis of methyl jasmonate and putative route to cis-jasmone*

INDUCTION OF DEFENCE BY *CIS*-JASMONE

Because of *cis*-jasmone's relationship with the jasmonate pathway, we decided to investigate whether aerial application of *cis*-jasmone could influence the defence of intact plants. This was achieved by placing low levels of *cis*-jasmone over bean plants contained in bell jars. The plants were tested for residual *cis*-jasmone, which was found to be completely absent after 24h. After a total of 48 h, these and control plants were placed in a wind tunnel and the effect on an aphid parasitoid, *Aphidius ervi*, was investigated. In both dual- and single-choice experiments, there were, respectively, threefold and twofold increases in oriented flight towards the *cis*-jasmone-treated plants, with both results being highly significant statistically (Birkett et al. 2000). One of the compounds showing induced release as a consequence of the *cis*-jasmone treatment was (*E*)-ocimene, which is known to be partly responsible for the response by *A. ervi* (Du et al. 1998). Although this compound was also induced by methyl jasmonate, the effect was short-lived and had disappeared 48 h after the initial treatment. However, the effect with *cis*-jasmone remained for 8 days (Birkett et al. 2000).

CIS-JASMONE AS AN ACTIVATOR OF GENE EXPRESSION

Bean plants treated with *cis*-jasmone, and also with methyl jasmonate as a positive control, were investigated by differential gene display. However, it was not possible to find genes for expressing the semiochemicals induced during the *cis*-jasmone treatment (Birkett et al. 2000). For example, the gene for (*E*)-ocimene synthase could not be located. It was therefore decided to change plants and to use the model plant *Arabidopsis thaliana*, for which there is full genomic sequence information and associated microarrays. Such microarrays were analysed (Matthes et al. 2003) by treatment with *cis*-jasmone and methyl jasmonate. A number of genes were found to be upregulated and this was confirmed by Northern blotting and other studies (Matthes et al. 2003). Currently, we are trying to use *A. thaliana* knockout mutants, over-expression in *A. thaliana* and heterologous expression in other systems to study, more easily, the biochemistry of gene products where the genes appear to be coding for enzymes that could be involved in defence or the persistent effect of *cis*-jasmone. For example, there are a number of cytochromes P450 and isoprenoid genes. In addition, there are genes that may be associated with the biosynthesis of *cis*-jasmone, and these include *OPR1* (Schaller et al. 2000; Schaller 2001) and a thiamine-diphosphate cofactor synthase gene (Vander Horn et al. 1993). It has also been possible to isolate the promoter sequence from some of these genes and to fuse this to the reporter luciferase, so that, when the plants are treated with *cis*-jasmone, this enzyme is produced and, on adding luciferin, the plants emit light. A considerable amount of work still needs to be done until we know how *cis*-jasmone is recognized by the plant and which genes are responsible for the long-term defence that we have found.

PRACTICAL USE OF *CIS*-JASMONE

Whilst we continue to investigate the molecular basis of *cis*-jasmone plant activation as a means of providing transgenic delivery of these types of crop protection approaches (Pickett and Poppy 2001), we have been looking at elite cereal cultivars for high levels of activation with *cis*-jasmone. A target for increased production is 6-methyl-5-hepten-2-one, one of a number of compounds (Quiroz et al. 1997) produced when *R. padi* attacks cereals and which causes repulsion of this aphid from normally attractive wheat seedlings. We also know that 6-methyl-5-hepten-2-one is an important foraging cue for the aphid parasitoid *A. ervi* (Du et al. 1998). The biosynthesis of 6-methyl-5-hepten-2-one has been reported (Demyttenaere and De Pooter 1996) as an oxidation product of isoprenoids by microbes. However, we have found that, in certain elite wheat cultivars, there is an upregulation of this compound with *cis*-jasmone. We also found that, as a consequence of this and other effects, there is repellency for the cereal aphid *S. avenae* in the olfactometer when the wheat cultivar has been treated with *cis*-jasmone (Bruce et al. 2003b). This has been followed through in the field where, in three seasons out of four, we have had statistically reduced levels of cereal aphids on winter wheat one month after *cis*-jasmone, as an emusifiable concentrate, was applied (Bruce et al. 2003b). Although we have been unable to do similar work on aphid parasitoids in the field, because of climatic problems, we have shown, in simulated field trials on wheat seedlings treated with *cis*-jasmone, that there is a statistically significantly increase in foraging by *A. ervi* (Bruce et al. 2003a).

SIGNALLING BY INTACT PLANTS

When barley plants are placed alongside certain weeds such as thistles (*Cirsium* spp.) in a convection-driven wind tunnel, they can become less attractive to aphids (Glinwood et al. 2004). Furthermore, it was shown that, if different cultivars of barley are similarly used as the 'inducing' and 'recipient' plants in such an experiment, then there can also be a reduction in aphid settling (Pettersson et al. 1999). Thus, when the cultivar Hulda was exposed to volatiles from another cultivar, Frida, the number of aphids settling was reduced by over 50%. Field trials showed a reduction of aphids on barley plants when intercropped with the appropriate 'inducing' cultivar. For example, there was a significant reduction of aphids on the cultivar Kara when it was grown in admixture in the field with the cultivars Frida and Alva (Ninkovic et al. 2002).

THE STIMULO-DETERRENT DIVERSIONARY STRATEGY, OR 'PUSH-PULL' STRATEGY

Although delivery of semiochemicals by plants, whether induced or not, provides a means of economically viable delivery, particularly for unstable or highly volatile compounds, the effects may not be sufficient to reduce the pest problem below the economic threshold (Chamberlain et al. 2001; Pickett et al. 2003). In an attempt to avoid rapid development of resistance to semiochemical control strategies, we and other groups put together a number of semiochemically based control methods into a stimulo-deterrent diversionary or 'push-pull' strategy (Miller and Cowles 1990; Smart et al. 1994; Pickett et al. 1997). This involves creating a 'push' effect from the main crop, using less attractive crop cultivars, repellents such as non-host volatiles, oviposition deterrent pheromones, or plant-derived antifeedants (see Figure 2).

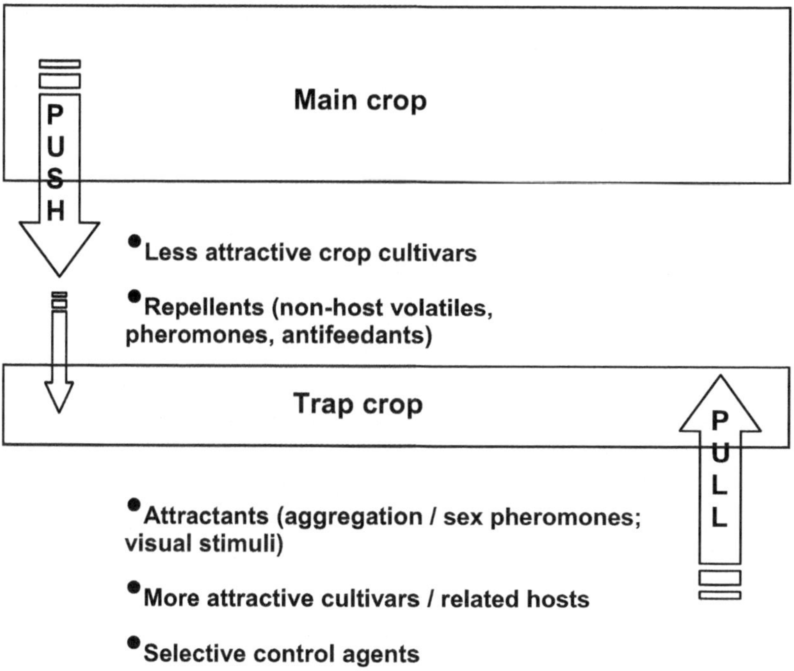

Figure 2. *The general principles of a 'push-pull' crop protection system*

This system also requires a trap crop ('pull') to which the pests are attracted by aggregation or sex pheromones, visual stimuli, or more attractive cultivars/related hosts. On the trap crop can also be deployed a highly selective control agent. Economics do not usually allow the use of biological control agents in broad-acre crops, but application to a limited area of trap crop, particularly one in which the best conditions for infectivity with the biological agent can be established, will make the process economically feasible. Into this system also comes the potential to exploit beneficial organisms such as predators and parasitoids of the pests and so, as part of the 'push' strategy, there is also an involvement of foraging cues to ensure that the main crop is visited by predators and parasitoids before the pest population builds up. We have attempted to do this in the U.K. on oilseed rape, initially using a trap crop comprising turnip rape, which produces both visual cues and volatile semiochemical attractants (Cook et al. 2004). Eventually, we hope to 'switch on' the effects of 'push' and 'pull' by means of plant activators such as *cis*-jasmone, described above.

PRACTICAL DEVELOPMENT OF A 'PUSH-PULL' AGAINST STEM-BORER MOTHS IN AFRICA

Working in collaboration with the International Centre of Insect Physiology and Ecology in Nairobi and its field station at Mbita Point, and with other agencies in Africa, including the Kenyan Agricultural Research Institute, we have helped to develop a system for controlling stem-borer moths, particularly in maize (Khan et al. 2000). Initially, alternative grass hosts were investigated by establishing a triplicated plot nursery at the field station at Mbita Point. The African colleagues made close observations of which grasses were favoured by the stem borers for oviposition and those which were not chosen. The main target pests were an indigenous noctuid, *Busseola fusca*, the maize stalk borer, and an introduced crambid, *Chilo partellus*, the spotted stalk borer. It was found that two forage grasses, Napier grass, *Pennisetum purpureum*, and Sudan grass, *Sorghum sudanensis*, were preferred to maize for oviposition by stem borers, and these plants were subsequently used as trap crops (the 'pull' effect) in field trials. Highly significant reductions of stem-borer numbers in maize were found when 50 m plots were surrounded by two or three rows of these trap crops and, in on-farm trials, yield increases of 1 to 1.5 tonnes per hectare were obtained (Khan et al. 2000). There was also a highly significant increase in oviposition in the trap crop as compared to the maize. In addition, Napier grass showed a low survival of the ensuing larvae, and it was found that a sticky secretion, produced within the stems by the presence of late larval instars, inundated the larvae and prevented their further development. Since the trap crop might be competitive with the maize, a gap was created between the trap crop and the main crop and, overall, there was a reduced area of the amount of maize produced. Therefore, any increase in yield as a consequence of stem-borer control needed to be set against control plots in which maize occupied the whole site.

Initially, the 'push' effect was created by one of the plants that was found not to be used for oviposition by stem borers, the molasses grass, *Melinis minutiflora*, also grown as a forage crop for cattle. This, planted between each row of maize, caused a dramatic reduction in stem borers (Khan et al. 2000), with a decrease in numbers of over 80%. Indeed, there was a highly significant reduction in stem borers at the more practically useful ratio of one row of *M. minutiflora* to three or four of maize. A statistically significant effect could still be seen at a ratio of one row in twenty rows of maize.

Using GC-EAG, we found key physiologically active compounds from the trap crops that were responsible for their high attractiveness to gravid stem-borer moths (Khan et al. 2000). We then turned to *M. minutiflora*, our hypothesis being that, as a non-host for these insects, there would be additional physiologically active compounds acting as repellents. This was indeed the case, and subsequent behavioural studies showed that the active compounds found specifically in *M. minutiflora*, but not in the trap-crop plants, comprised (*E*)-ocimene, (*E*)-4,8-dimethyl-1,3,7-nonatriene, (-)-β-caryophyllene, humulene and α-terpinolene (Figure 3). On noting the presence of the first two compounds, we realized that *M. minutiflora* is treated as a non-host because it produces chemicals that would be emitted by a highly infested maize plant. We subsequently showed that this phenomenon was responsible for the increased foraging by parasitoids of the stem borers (Khan et al. 1997). For example, in the Y-tube olfactometer, the parasitoid *Cotesia sesamiae* responded to the nonatriene at a similar level to that found in the live plant and in an extract of the plant. Indeed, in two of the trial areas, one near Mbita Point in Nyanza Province, Suba District, and the other in the high maize-yielding area near Kitale in Trans Nzoia, use of one row of *M. minutiflora* to three rows of maize gave highly significant increases in foraging by stem-borer parasitoids.

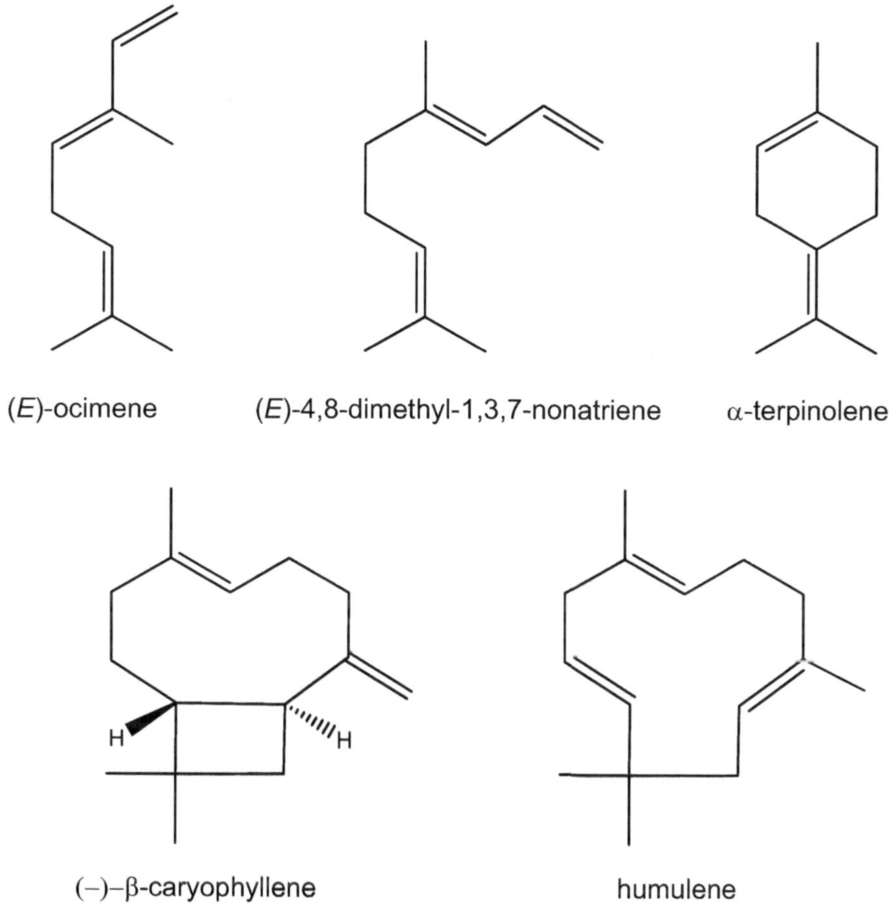

Figure 3. *Electrophysiologically active compounds identified in Melinis minutiflora volatiles*

AFRICAN STEM BORER 'PUSH-PULL' CONTROL – THE WAY FORWARD

We would like to review the prospect of using biotechnological approaches to maximizing and exploiting these effects on stem borers. We could contemplate transferring the systemic release of the nonatriene from *M. minutiflora* to maize itself. However, it must be remembered that maize under insect attack, as referred to above, already produces the nonatriene. What we really require is a maize plant that produces the nonatriene by induction more effectively. Often, the laying of eggs can induce defence (Blaakmeer et al. 1994; Hilker et al. 2002; Hilker and Meiners 2002), so if we could, even by conventional plant breeding, enhance the response of the plant to egg-laying in terms of nonatriene production, then this would give an early defence against colonization of maize by stem borers and may remove the need for the laborious intercropping approach. Nonetheless, it must be pointed out that, once the intercrop has been established, then the farmer only has to keep the plot free of extraneous and aggressive plant material and the system will largely look after itself. It will produce not only a higher maize yield, even taking into account the smaller area through the loss of land to the trap crop, but will also have added value in terms of the cattle forage provided by both the trap crop and the intercrop (Khan and Pickett 2004). Indeed, involvement with farmers, particularly at Farmers' Days (barazas), has introduced a number of ideas and an alternative for the term 'push-pull', which in Kiswahili is reversed to 'pull-push', or 'vuta sukuma'.

We have had requests that we should use, as an intercrop, a legume rather than a grass. The farmers would very much like to grow edible legumes. We have, as yet, been unable to find an edible legume that has the effect of attracting stem-borer parasitoids. Nonetheless, a series of forage legumes in the *Desmodium* genus such as silverleaf desmodium, *D. uncinatum*, do repel stem borers when used in the intercropping system. However, during these trials, we noticed, with great surprise, that the desmodium was also controlling another extremely important pest in sub-Saharan subsistence agriculture, the African witchweed, *Striga hermonthica* (Scrophulariaceae) (Matúšová and Bouwmeester in press). The pernicious striga weed develops underground as a parasite on the maize roots and then appears above the surface, where it begins to photosynthesize and produces beautiful purple flowers, setting seed which will remain viable in the soil for up to 20 years.

Striga has received a considerable amount of attention, but most of the really effective solutions involve more expensive technology than is normally available to subsistence farmers in these circumstances. However, with the *Desmodium* intercrop, there is a tremendous impact on striga development. We have subsequently shown that this is through a suicidal germination mechanism in which allelopathic chemicals are produced by the desmodium roots, some causing a dramatic germination of striga seeds, but others preventing the development of the subterranean phase of the parasite and thereby inhibiting colonization of the maize plant (Khan et al. 2002; Tsanuo et al. 2003). There has been a rapid take-up of this approach by farmers, and we have used various media instruments for promoting this, including pamphlets and a regular radio programme. The farmers themselves have transferred the technology at Farmers' Days and also, in one district, Vihiga, by putting on an extremely innovative show about the whole 'push-pull' system. There are now 15 regions using the 'push-pull' approach, involving over 4,000 farmers in many of the regions around the Victoria Lake basin, originally starting in Kenya but now including Uganda and Tanzania (Khan and Pickett 2004). When tested in comparative trials, this approach has proved to be more effective than use of pesticides, and substantially cheaper (Parrott 2005).

CONCLUSIONS

Thus, it can be seen that understanding the interactions of plants with insects can yield new ways of exploiting, at the practical level, plant defence. This may be delivered by application of natural plant activators or intercropping regimes and a 'push-pull' system. Basic science, and particularly understanding the chemical ecology of pest–plant interaction by combined analytical-chemical, neurophysiological and behavioural studies, can lead through to real practical developments.

ACKNOWLEDGEMENTS

Rothamsted Research receives grant-aided support from the Biotechnology and Biological Sciences Research Council (BBSRC), UK, with additional funding provided under the Biological Interactions in the Root Environment (BIRE) initiative. Field studies in Africa are funded by the Gatsby Charitable Foundation and the Rockefeller Foundation. This work was in part supported by the Department for Environment, Food and Rural Affairs (Defra), UK.

REFERENCES

Aharoni, A., Giri, A.P., Deuerlein, S., et al. 2003. Terpenoid metabolism in wild-type and transgenic *Arabidopsis* plants. *Plant Cell,* 15 (12), 2866-2884.

Beale, M.H., Birkett, M.A., Bruce, T.J.A., et al. in prep. Aphid alarm pheromone from transgenic plants affects aphids and their parasitoids.

Birkett, M.A., Campbell, C.A.M., Chamberlain, K., et al. 2000. New roles for *cis*-jasmone as an insect semiochemical and in plant defense. *Proceedings of the National Academy of Sciences of the United States of America,* 97 (16), 9329-9334.

Blaakmeer, A., Hagenbeek, D., Van Beek, T.A., et al. 1994. Plant response to eggs vs. host marking pheromone as factors inhibiting oviposition by *Pieris brassicae*. *Journal of Chemical Ecology,* 20 (7), 1657-1665.

Bruce, T.J., Pickett, J.A. and Smart, L.E., 2003a. *cis*-Jasmone switches on plant defence against insects. *Pesticide Outlook,* 14 (3), 96-98.

Bruce, T.J.A., Martin, J.L., Pickett, J.A., et al. 2003b. *cis*-Jasmone treatment induces resistance in wheat plants against the grain aphid, *Sitobion avenae* (Fabricius) (Homoptera: Aphididae). *Pest Management Science,* 59 (9), 1031-1036.

Bruce, T.J.A., Wadhams, L.J. and Woodcock, C.M., 2005. Insect host location: a volatile situation. *Trends in Plant Science,* 10 (6), 269-274.

Chamberlain, K., Guerrieri, E., Pennacchio, F., et al. 2001. Can aphid-induced plant signals be transmitted aerially and through the rhizosphere? *Biochemical Systematics and Ecology,* 29 (10), 1063-1074.

Cook, S.M., Watts, N.P., Hunter, F., et al. 2004. Effects of a turnip rape trap crop on the spatial distribution of *Meligethes aeneus* and *Ceutorhynchus assimilis* in oilseed rape. *IOBC/WPRS Bulletin,* 27 (10), 199-206.

Demyttenaere, J.C.R. and De Pooter, H.L., 1996. Biotransformation of geraniol and nerol by spores of *Penicillium italicum. Phytochemistry Oxford,* 41 (4), 1079-1082.

Dicke, M. and Sabelis, M.W., 1988. Infochemical terminology: based on cost-benefit analysis rather than origin of compounds? *Functional Ecology,* 2 (2), 131-139.

Doughty, K.J., Kiddle, G.A., Pye, B.J., et al. 1995. Selective induction of glucosinolates in oilseed rape leaves by methyl jasmonate. *Phytochemistry,* 38 (2), 347-350.

Du, Y.J., Poppy, G.M., Powell, W., et al. 1998. Identification of semiochemicals released during aphid feeding that attract parasitoid *Aphidius ervi*. *Journal of Chemical Ecology,* 24 (8), 1355-1368.

Farmer, E.E. and Ryan, C.A., 1990. Interplant communication airborne methyl jasmonate induces synthesis of proteinase inhibitors in plant leaves. *Proceedings of the National Academy of Sciences of the United States of America,* 87 (19), 7713-7716.

Glinwood, R., Ninkovic, V., Pettersson, J., et al. 2004. Barley exposed to aerial allelopathy from thistles (*Cirsium* spp.) becomes less acceptable to aphids. *Ecological Entomology,* 29 (2), 188-195.

Hardie, J., Isaacs, R., Pickett, J.A., et al. 1994. Methyl salicylate and (-)-(1*R*,5*S*)-myrtenal are plant-derived repellents for black bean aphid, *Aphis fabae* Scop. (Homoptera: Aphididae). *Journal of Chemical Ecology,* 20 (11), 2847-2855.

Hilker, M. and Meiners, T., 2002. Induction of plant responses to oviposition and feeding by herbivorous arthropods: a comparison. *Entomologia Experimentalis et Applicata,* 104 (1), 181-192.

Hilker, M., Rohfritsch, O. and Meiners, T., 2002. The plant's response towards insect egg depostion. *In:* Hilker, M. and Meiners, T. eds. *Chemoecology of insect eggs and egg deposition*. Blackwell Publishers, Berlin, 205-233.

Karban, R., Baldwin, I.T., Baxter, K.J., et al. 2000. Communication between plants: induced resistance in wild tobacco plants following clipping of neighboring sagebrush. *Oecologia,* 125 (1), 66-71.

Khan, Z.R., Ampong-Nyarko, K., Chiliswa, P., et al. 1997. Intercropping increases parasitism of pests. *Nature,* 388 (6643), 631-632.

Khan, Z.R., Hassanali, A., Overholt, W., et al. 2002. Control of witchweed *Striga hermonthica* by intercropping with *Desmodium* spp., and the mechanism defined as allelopathic. *Journal of Chemical Ecology,* 28 (9), 1871-1885.

Khan, Z.R. and Pickett, J.A., 2004. The 'push-pull' strategy for stemborer management: a case study in exploiting biodiversity and chemical ecology. *In:* Gurr, G.M., Wratten, S.D. and Altieri, M.A. eds. *Ecological engineering for pest management: advances in habitat manipulation for arthropods*. CSIRO, Collingwood, 155-164.

Khan, Z.R., Pickett, J.A., Van den Berg, J., et al. 2000. Exploiting chemical ecology and species diversity: stem borer and striga control for maize and sorghum in Africa. *Pest Management Science,* 56 (11), 957-962.

Koch, T., Bandemer, K. and Boland, W., 1997. Biosynthesis of *cis*-jasmone: a pathway for the inactivation and the disposal of the plant stress hormone jasmonic acid to the gas phase? *Helvetica Chimica Acta,* 80 (3), 838-850.

Lucas, J.A., 1999. Plant immunisation: from myth to SAR. *Pesticide Science,* 55 (2), 193-196.

Matthes, M., Napier, J.A., Pickett, J.A., et al. 2003. New chemical signals in plant protection against herbivores and weeds. *In: The BCPC international congress Crop science & technology, Glasgow, 10-12 November 2003*. British Crop Protection Council, Alton, 1227-1236.

Matúšová, R. and Bouwmeester, H.J., in press. The effect of host-root-derived chemical signals on the germination of parasitic plants. *In:* Dicke, M. and Takken, W. eds. *Chemical ecology: from gene to ecosystem*. Springer, Dordrecht. Wageningen UR Frontis Series no. 16. [http://library.wur.nl/frontis/chemical_ecology/04_matusova.pdf]

Miller, J.R. and Cowles, R.S., 1990. Stimulo-deterrent diversion: a concept and its possible application to onion maggot control. *Journal of Chemical Ecology,* 16 (11), 3197-3212.

Ninkovic, V., Olsson, U. and Pettersson, J., 2002. Mixing barley cultivars affects aphid host plant acceptance in field experiments. *Entomologia Experimentalis et Applicata,* 102 (2), 177-182.

Nordlund, D.A. and Lewis, W.J., 1976. Terminology of chemical releasing stimuli in intraspecific and interspecific interactions. *Journal of Chemical Ecology,* 2 (2), 211-220.

Parrott, S., 2005. *The quiet revolution: push-pull technology and the African farmer*. Gatsby Charitable Foundation, London. Gatsby Occasional Paper.

Pettersson, J., Ninkovic, V. and Ahmed, E., 1999. Volatiles from different barley cultivars affect aphid acceptance of neighbouring plants. *Acta Agriculturae Scandinavica. Section B. Soil and Plant Science,* 49 (3), 152-157.

Pettersson, J., Pickett, J.A., Pye, B.J., et al. 1994. Winter host component reduces colonization by bird-cherry-oat aphid, *Rhopalosiphum padi* (L.) (Homoptera, Aphididae), an other aphids in cereal fields. *Journal of Chemical Ecology,* 20 (10), 2565-2574.

Pickett, J.A. and Poppy, G.M., 2001. Switching on plant genes by external chemical signals. *Trends in Plant Science,* 6 (4), 137-139.

Pickett, J.A., Rasmussen, H.B., Woodcock, C.M., et al. 2003. Plant stress signalling: understanding and exploiting plant-plant interactions. *Biochemical Society Transactions,* 31 (1), 123-127.

Pickett, J.A., Wadhams, L.J. and Woodcock, C.M., 1997. Developing sustainable pest control from chemical ecology. *Agriculture, Ecosystems and Environment,* 64 (2), 149-156.

Pickett, J.A., Wadhams, L.J., Woodcock, C.M., et al. 1992. The chemical ecology of aphids. *Annual Review of Entomology,* 37, 67-90.

Preston, C.A., Betts, H. and Baldwin, I.T., 2002. Methyl jasmonate as an allelopathic agent: sagebrush inhibits germination of a neighboring tobacco, *Nicotiana attenuata*. *Journal of Chemical Ecology,* 28 (11), 2343-2369.

Quiroz, A., Pettersson, J., Pickett, J.A., et al. 1997. Semiochemicals mediating spacing behavior of bird cherry-oat aphid, *Rhopalosiphum padi* feeding on cereals. *Journal of Chemical Ecology,* 23 (11), 2599-2607.

Schaller, F., 2001. Enzymes of the biosynthesis of octadecanoid-derived signalling molecules. *Journal of Experimental Botany,* 52 (354), 11-23.

Schaller, F., Biesgen, C., Müssig, C., et al. 2000. 12-Oxophytodienoate reductase 3 (OPR3) is the isoenzyme involved in jasmonate biosynthesis. *Planta,* 210 (6), 979-984.

Shulaev, V., Silverman, P. and Raskin, I., 1997. Airborne signalling by methyl salicylate in plant pathogen resistance. *Nature,* 385 (6618), 718-721.

Smart, L.E., Blight, M.M., Pickett, J.A., et al. 1994. Development of field strategies incorporating semiochemicals for the control of the pea and bean weevil, *Sitona lineatus* L. *Crop Protection,* 13 (2), 127-135.

Tsanuo, M.K., Hassanali, A., Hooper, A.M., et al. 2003. Isoflavanones from the allelopathic aqueous root exudate of *Desmodium uncinatum*. *Phytochemistry,* 64 (1), 265-273.

Vander Horn, P.B., Backstrom, A.D., Stewart, V., et al. 1993. Structural genes for thiamine biosynthetic enzymes (thiCEFGH) in *Escherichia coli* K-12. *Journal of Bacteriology,* 175 (4), 982-992.

CHAPTER 12

CHEMICAL ECOLOGY FROM GENES TO COMMUNITIES

Integrating 'omics' with community ecology

MARCEL DICKE

Laboratory of Entomology, Wageningen University and Research Centre, P.O. Box 8031, 6700 EH Wageningen, The Netherlands. http://www.insect-wur.nl.
E-mail: Marcel.Dicke@wur.nl

Abstract. Chemical cues that convey information are widely used by living organisms. The cues mediate interactions in food webs as well as non-trophic interactions such as interactions between conspecific organisms or between plants and natural enemies of herbivorous organisms. Communities are composed of food webs and each food web is overlaid with a reticulate infochemical web that is more complex than the underlying food web. Chemical ecology has addressed the role of information conveyance in intraspecific and interspecific interactions and has mostly concentrated on elucidating the identity of chemicals and their role in individual interactions of food webs. In addition, the role of infochemicals has been investigated in multitrophic interactions.

Recently, several exciting developments have taken place. On the one hand, chemical ecologists more and more address molecular mechanisms underlying the production of infochemicals and the responses to the cues, such as signal transduction and gene expression. On the other hand, studies on the role of infochemicals in population and community ecology have been initiated. These developments are not independent of each other, and knowledge of mechanisms will provide important tools for investigating the role of infochemicals in populations and communities. This will be discussed especially in the context of insect–plant communities.
Keywords: ecogenomics; phenomics; herbivore-induced plant volatiles; signal transduction; community ecology; behavioural ecology

INTRODUCTION

Chemical information conveyance is omnipresent in biological systems. Chemical cues are a major source of information for very different organisms ranging from micro-organisms to mammals (e.g., Dicke and Grostal 2001; Kats and Dill 1998; Penn 2002; Roitberg and Isman 1992; Tollrian and Harvell 1999), and infochemicals play a role in terrestrial, aquatic and soil ecosystems (Van Tol et al. 2001; Rasmann et al. 2005; Roitberg and Isman 1992; Tollrian and Harvell 1999; Dicke and Takken in press). Chemical information affects various behaviours that underlie population dynamics and food-web interactions, including the selection of food, the selection of

M. Dicke and W. Takken (eds.), Chemical Ecology: From Gene to Ecosystem, 175-189.

mates, competition and the avoidance of predators (e.g., Dicke and Vet 1999; Hilker et al. 2002; Kats and Dill 1998; Roitberg and Isman 1992; Sabelis et al. 1999; Turlings and Benrey 1998; Wertheim et al. 2005). Therefore, chemical information is an important factor influencing species interactions and most likely also community processes (Kessler et al. 2004; Van Donk in press; Vet 1999). However, the study of chemical information conveyance has been mostly restricted to studies at the level of individual organisms and the identification of the chemicals that convey the information. The influence of chemical information on food-web processes has received little attention (Hunter 2002; Vet 1999; Wertheim et al. 2005), in contrast to effects of direct trophic interactions (Morin 1999). Yet, circumstantial evidence indicates that chemical information from phenotypically plastic plants can have important influences on food-web dynamics through indirect effects that combine bottom-up and top-down effects (Dicke and Vet 1999; Sabelis et al. 1999). Moreover, pheromones that mediate intraspecific interactions among animals may have important consequences for food-web interactions (Wertheim 2005; Wertheim et al. in press). Empirical support should come from manipulative experiments, such as manipulations of infochemical emission phenotype, that compare food-web processes in the presence and absence of infochemicals. The ability to manipulate the infochemical phenotype in specific and well-known ways is indispensable for this approach. Such experiments have recently come within reach. This provides a modern, novel and exciting interdisciplinary approach to ecology that is possible because of recent breakthroughs at the level of subcellular processes (e.g., Baldwin et al. 2001; Dicke et al. 2004; Dicke and Van Poecke 2002; Fitzpatrick et al. 2005; Jacobs et al. 2005; Kessler et al. 2004; Mitchell-Olds 2001; Van Poecke and Dicke 2004), in metabolomic approaches (Fiehn 2002) and in quantitative food-web analysis (Omacini et al. 2001; Rott and Godfray 2000). For the design of manipulative experiments and for understanding their outcome, information on mechanisms underlying ecological processes is essential (Dicke et al. 2004; Kessler et al. 2004; Wertheim et al. 2005). In this paper I will address the potential of interdisciplinary approaches to the unravelling of the role of chemical information conveyance in community processes.

COMMUNITIES AND FOOD WEBS

Communities are complex compositions of hundreds of interacting species at different trophic levels (Morin 1999). One way of appreciating the complexity of communities is to analyse food webs. This shows that, for example, even a small part of an insect food web may comprise tens of species (Figure 1). Rott and Godfray sampled *Phyllonorycter* leaf-miner moths and their parasitoids on four plant species in a 10,000 m^2 area (Rott and Godfray 2000). They recorded twelve *Phyllonorycter* species and 27 of their parasitoids. The composition of the food web changed in time. Some species that were abundant in the summer were scarce in autumn and vice versa. The estimated total number of *Phyllonorycter* ranged from a few million up to 75 million individuals. Given that the leaf miners are also attacked by predators and pathogens and that the host plants are also attacked by dozens of other herbivore species it is clear that food webs are highly complex in composition and that this complexity is variable in space and time.

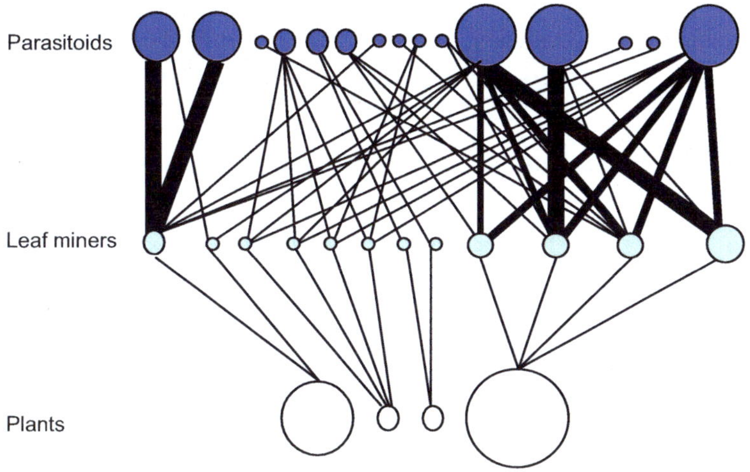

Figure 1. *Schematic representation of a food web consisting of 4 plant species, 12 herbivorous leaf miners and 27 parasitoids. Relative sizes of circles within a trophic level indicate relative population sizes. Lines indicate trophic interactions. Relative thickness of lines between parasitoid species and leaf-miner species indicate relative number of parasitoids involved in the interaction. Figure based on data in Rott and Godfray (2000)*

Food-web analyses address interactions within communities that reflect direct trophic relationships, where one organism feeds on another. Yet, although these interactions are important in shaping communities, they reflect only a part of the interactions in a community. Members of communities also show indirect interactions, i.e., interactions between two organisms that are mediated by a third organism, that connect organisms that do not have a trophic relationship. Such indirect interactions may be important in shaping communities as well.

Major recent developments in our understanding of communities are that:

1. indirect interactions have important effects on food web dynamics (e.g., Abrams et al. 1996; Bonsall and Hassell 1997);
2. top-down and bottom-up forces (enemy-controlled versus resource-controlled forces, respectively) are often integrated rather than mutually exclusive: plants can dramatically influence top-down forces on herbivores, as mediated by carnivorous enemies of the herbivores (Bernays 1998; Dicke and Vet 1999; Sabelis et al. 1999);
3. species characteristics appear to be phenotypically plastic and consequently the effects of species on interactions in a food web are dynamic (Agrawal 2001); ecogenomics can link phenotype to genotypic expression (Baldwin et al. 2001; Dicke et al. 2004; Kessler et al. 2004).

Both direct and indirect interactions are mediated by infochemicals. Communities and food webs are, therefore, overlaid with infochemical webs.

INFOCHEMICAL WEBS

Given that each organism in a community emits and responds to chemical information, it is clear that communities abound with interactions mediated by chemical information. Although chemical information by itself cannot be used to build bodies, it essentially influences interactions between organisms and, thus, fitness of individuals, and most likely also food web and community processes (Dicke and Hilker 2003).

Infochemically mediated interactions can in principle occur between any two organisms in a community, whether conspecific or heterospecific, whether connected by a food-web interaction or not (Dicke and Sabelis 1992; Stowe et al. 1995). For instance, conspecific organisms may interact through pheromones, and these pheromones may be exploited by their natural enemies. E.g., male *Pieris brassicae* butterflies endow a female with an anti-aphrodisiac pheromone during mating. This pheromone renders the females less attractive to other males that might compete with the original male for offspring. However, in addition to the intraspecific interaction, the pheromone also mediates an indirect interaction. The egg parasitoid *Trichogramma brassicae* is attracted to the anti-aphrodisiac pheromone, and after arrival at the mated female butterfly, the wasp mounts the butterfly and hitches a ride to the spot where her transporter deposits her eggs. These eggs are subsequently parasitized by the *Trichogramma* wasp (Fatouros et al. 2005). Thus, an infochemical mediating a non-trophic interaction (mating) can also mediate a trophic interaction (parasitization). Because a single infochemical may mediate many interactions in a food web, the costs and benefits of an infochemical to an emitting organism include the costs and benefits of each of these interactions. Therefore, the evolutionary ecology of infochemicals can only be understood in a community context (Dicke and Sabelis 1992).

Indirect interactions that do not involve trophic relationships are, e.g., those between plants and carnivorous arthropods such as predators and parasitoids of herbivores. In response to herbivory an individual plant produces a complex blend of volatiles. As a result, enemies of the herbivore are attracted to the plant. This is a general phenomenon that has been recorded for a large number of plants (Dicke 1999; 2000; Hilker and Meiners 2002; Turlings et al. 1993). Just as in the case of *Pieris*'s anti-aphrodisiac pheromone, also herbivore-induced plant volatiles can be exploited by many other organisms in a community, including, e.g., herbivores and neighbouring plants (Dicke and Bruin 2001; Dicke and Van Loon 2000; Hilker and Meiners 2002) (Figure 2).

Thus, a food web is overlaid with an infochemical web. Moreover, because each infochemical may mediate many interactions, both trophic interactions and indirect, non-trophic, interactions, the infochemical web is more complex than the food web. Therefore, when investigating communities, infochemical webs should be studied in addition to food webs.

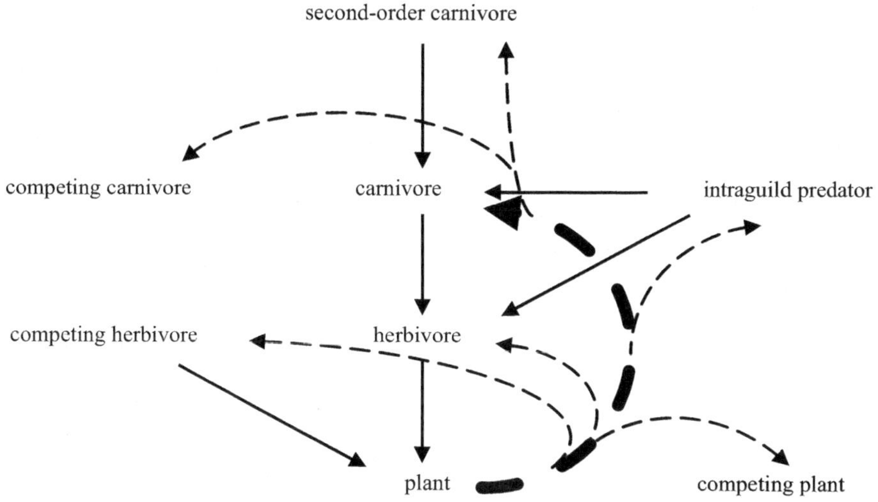

Figure 2. *Herbivore-induced plant volatiles are well-known to attract carnivorous arthropods that feed on the inducing herbivore. Moreover, they may affect the behaviour of many other organisms in a food web, including other plants, herbivores, carnivores and second-order carnivores. Solid lines indicate trophic interactions (food web) and broken lines indicate interactions influenced by herbivore-induced plant volatiles. Note that only one single infochemical blend, viz., the herbivore-induced plant volatiles, is depicted. Each component of the food web may emit infochemicals that similarly affect various other players in the food web*

MECHANISMS OF CHEMICAL INFORMATION CONVEYANCE

Before an infochemical can influence interactions between individual organisms and consequently food webs and community processes, a range of processes has been initiated, from gene expression to mechanisms of storing and releasing the compounds. It is well-known that organisms regulate the production and emission of infochemicals. After all, both the production and the emission come with costs and benefits and organisms are under selection to maximize the returns of infochemical emission. Understanding the mechanisms is essential for understanding the expression of phenotypes in terms of infochemical emission. Moreover, understanding these mechanisms also allows the careful manipulation of infochemical-emission phenotypes (Dicke et al. 2003; Kessler and Baldwin 2004; Kessler et al. 2004; Van Poecke and Dicke 2002), and therefore provides ecologists with important new tools to investigate the ecology of chemical information conveyance.

Apart from variation in infochemical emission, the response to infochemicals may vary as well. Animals are well-known to be phenotypically plastic in their responses to infochemicals (Papaj and Lewis 1993). Moreover, knowledge on genes involved in behavioural responses of animals is rapidly accumulating and this provides exciting new opportunities for manipulative experiments (Fitzpatrick et al. 2005). For instance, the perception of infochemicals, their decoding in the brain and memory processes involved are all processes that are intensively investigated at the molecular level (Fitzpatrick et al. 2005).

At the mechanistic level, systems are characterized by complexity, just as is the case at higher levels of integration (Figure 3). For instance, odour blends are complex mixtures ranging from a few to hundreds of components (Roitberg and Isman 1992; Turlings et al. 1993; Van den Boom et al. 2004), the metabolome of organisms is highly complex with developmental and temporal variation (Fiehn 2002; Rosenthal and Berenbaum 1992), and transcriptomic changes are substantial and dynamic and highly variable with developmental and environmental conditions (De Vos et al. 2005; Heidel and Baldwin 2004; Reymond et al. 2004; Schenk et al. 2000). Thus, connecting ecology with mechanisms is very much a matter of dealing with complexity at different levels. This means that intelligent decisions have to be made at different levels of biological organization so as to unravel the patterns shaping these complex biological systems.

In the remainder of this chapter I will review how knowledge of mechanisms can be exploited to develop new strategies to understanding the effects of chemical signalling on communities.

Genome	10,000s of genes
Transcriptome	1000s to 10,000s of mRNAs
Proteome	1000s to 10,000s of proteins
Metabolome	10,000s of metabolites
Phenome	1000s of phenotypes
Infochemicals	Few to 10s of components in each infochemical; each organism in a community emits infochemicals
Community	1000s of interacting species, each with variable phenotypes

Figure 3. *Degree of complexity at different levels of biological organization, from the genome to the community*

COMMUNITY APPROACH

To understand how chemical information influences community processes it is essential to take an experimental, manipulative approach. Ecologists are well aware of the value of a comparative approach where individuals with different phenotypes are compared. However, it is not always easy to manipulate a phenotype in a biologically realistic way, especially in relation to its infochemical emission or perception. One may apply a synthetic infochemical, and this has been extensively done. However, this option is limited to situations where the infochemical's composition is relatively simple and the components can be synthesized or extracted in pure form. However, in many cases the composition of an infochemical is complex and the components may comprise stereochemically active compounds that are difficult to synthesize *in vitro*. Moreover, if the infochemical emission is phenotypically plastic or temporally dynamic, this may be difficult to mimic realistically. For instance, moths emit a sex pheromone during a restricted period during the night and so the application of a synthetic pheromone in a trap in the environment may be useful for investigating which species respond to the pheromone but it may be less suited for investigating the effect of the pheromone on community processes.

Several approaches have recently been taken to investigate the effect of chemical information on community processes. These manipulative approaches strongly depend on mechanistic information

Drosophila aggregation pheromone

One approach is to distribute the infochemical in synthetic form in a community. Drosophilid fruit flies aggregate on food and oviposition substrates and this behaviour is mediated by an aggregation pheromone (Bartelt et al. 1985; Wertheim et al. in press). Males produce the pheromone and transfer it to the female during mating. The females deposit the pheromone on the oviposition substrate during egg deposition. The emission from the substrate continues for at least several days. The aggregation pheromones of *Drosophila melanogaster* and *D. simulans* consist of a single compound, viz., cis-vaccenyl acetate, which is available in synthetic form. The synthetic pheromone or the naturally deposited pheromone can be used rather easily in manipulative field experiments (Wertheim et al. in press). Laboratory experiments have demonstrated that parasitoids of *Drosophila* larvae exploit this aggregation pheromone and are attracted to the site of oviposition (Hedlund et al. 1996). The application of the pheromone on substrates or fruits in an orchard resulted in the attraction of *Drosophila* flies whose pheromone was applied, as well as the attraction of other, competing *Drosophila* species. The degree of attraction was dose-dependent. In addition, parasitoids of *Drosophila* larvae were attracted (Wertheim et al. in press). Thus, the application of the synthetic pheromone in the field can provide information on its effects on community members and their aggregation. The consequences of the presence of the *Drosophila* aggregation pheromone were both direct effects (attraction of conspecifics and heterospecifics) and indirect effects (for instance, increased interference as a result of higher densities) (Wertheim et al. in press). A similar approach has also been taken for investigating the effects of herbivore-induced plant volatiles (Kessler and Baldwin 2001). This approach will provide the best information if the emission dynamics of the synthetic infochemical sufficiently resemble the natural emission dynamics.

Phytohormonal induction of plant volatiles

Another approach is to manipulate an individual's phenotype by manipulating natural biosynthetic pathways. The phytohormone jasmonic acid is well-known to mediate many phenotypic changes in plants (Dicke and Van Poecke 2002; Kessler et al. 2004). The application of jasmonic acid to tomato plants under field conditions resulted in an increased attraction of parasitic wasps to caterpillar-damaged tomato plants (Thaler 1999) and jasmonic-acid application was found to influence the composition of the insect community on the plants (Thaler 2002; Thaler et al. 2001). However, the disadvantage of this approach is that the single application of an external jasmonic-acid dose is likely to be very different from the natural induction dynamics and concentration. Yet, this approach may provide interesting information, as was clearly shown for the tomato studies (Thaler 1999; 2002; Thaler et al. 2001).

Molecular-genetic approach to plant characteristics

A third approach is to compare two genotypes that differ in identified components of their genetic background. This can relate to a well-characterized mutant and the related wildtype genotype or to a transgenic plant in which one gene is overexpressed or knocked out versus its wildtype. This approach has been taken for the wild tobacco plant *Nicotiana attenuata* (Kessler in press; Kessler et al. 2004). Kessler et al. (2004) knocked out three genes from the jasmonate signal transduction pathway. This pathway leads to the production of jasmonic acid or fatty-acid-derived green-leaf odours. Jasmonic acid is well-known to be involved in induced defences of plants against herbivorous insects, and green-leaf odours are known to influence behaviour of herbivorous insects and their natural enemies (Dicke and Van Poecke 2002). By generating three plant lines, each knocked out for one gene of the jasmonate pathway and bringing these plants into their natural environment, Kessler and colleagues investigated the effects of the genetic modification on herbivory. When planted into their native habitat, lipoxygenase-deficient plants were more vulnerable to *N. attenuata*'s adapted herbivores but were also exploited by a herbivore species that was otherwise not found on *N. attenuata*, which fed and reproduced successfully on the *LOX 3*-deficient plants. In addition to observing changes in the insect community as a result of transforming the plants, Kessler and co-workers also assessed the effects at the transcriptomic level through a dedicated microarray analysis and at the level of volatile emission through GC-MS analysis of the headspace of transformed plants (Kessler et al. 2004). This approach is essentially dependent on knowledge of the mechanism underlying induced plant defence, such as the involvement of signal transduction pathways, essential genes in the pathways and the function of their products, as well as knowledge of manipulating the expressed genotype by, e.g., anti-sense knock out, virus-induced gene silencing or RNA interference (Hamilton and Baulcombe 1999; Robertson 2004). Their approach is a major step forward in understanding the role of certain genes and their products in the effects on community ecology. In this case, the effect of the presence/absence of activity of a single gene was compared. Progress in molecular biology is likely to yield other, even more exciting, tools for ecologists as well, including those that allow for quantitative differences (Dicke et al. 2004). This is likely to bring unprecedented progress in our understanding of the role of infochemicals in community processes.

Complementing field studies with laboratory studies

The information from field studies needs to be complemented by laboratory studies to elucidate whether, where and to what extent infochemicals influence the outcome of individual interactions. In the laboratory one can disentangle the complexity of interactions in a directed way by investigating those interactions that are most likely to be influenced. Laboratory information on the effects of individual genes on infochemically mediated interactions is still scarce. For example, potato plants that had been transformed with a linalool synthase gene from strawberry behind a 35S promoter constitutively emitted the monoterpene linalool. These transgenic potato plants attracted the predatory mite *Phytoseiulus persimilis* (Bouwmeester et al. 2003), which is known to be attracted to synthetic linalool (Dicke et al. 1990). Linalool is one of the volatiles induced in lima-bean plants by feeding damage inflicted by the spider mite *Tetranychus urticae,* the prey of *P. persimilis* (Dicke et al. 1990). The parasitoid *Cotesia rubecula* is attracted to volatiles from *Arabidopsis thaliana* that is infested with caterpillars of *Pieris rapae* (Van Poecke et al. 2001). However, the attraction was impaired when Arabidopsis plants were used in which the *LOX 3* gene was co-suppressed, which blocks induction of jasmonic acid, or in plants in which a bacterial *NahG* gene was inserted, which results in breakdown of salicylic acid (Van Poecke and Dicke 2002). Moreover, in tomato the attraction of the predatory mite *P. persimilis* to herbivore-damaged plants was impaired in plants with a mutation in the jasmonic-acid signal transduction pathway compared to wildtype plants (Ament et al. 2004; Thaler et al. 2002). These studies show that single biosynthetic genes or genes that interfere with signal transduction pathways can influence infochemically mediated interactions between plants, herbivorous arthropods and their natural enemies.

'OMICS' AND COMMUNITY ECOLOGY

In the interaction of organisms with their environment, each individual expresses a complex phenotype that is subject to plasticity in response to the environment or in response to the individual's phenology (Agrawal 2001). In fact, the phenotype is not a static but a highly dynamic feature. The changes may occur over different spatial scales (from the organelle to the organ) and over different temporal scales (from milliseconds to days or longer). Moreover, the phenotype is influenced by many genetic components. To investigate the contribution of individual traits, one should ideally manipulate that trait so as to affect its expression in the natural – though complex – way. The best way of doing this is to use mutants that are altered in the expression of the trait (Dicke et al. 2004; Dicke and Van Poecke 2002; Kessler et al. 2004; Roda and Baldwin 2003). In fact, comparing mutants with their relevant wildtype allows one to analyse the effects of genetic variation in single traits, and thus to assess the role of these genes in the species' ecology.

To date, a molecular-genetic approach to chemical ecology and community ecology is rapidly developing (Baldwin 2001; Dicke et al. 2004; Kessler et al. 2004). In order to take this novel approach, one needs to have a suitable system that provides all necessary tools. In the past few years we have developed *Arabidopsis thaliana* as a model for a molecular-genetic approach of the ecology of herbivore-induced plant volatiles (e.g., Van Poecke and Dicke 2002; 2003; Van Poecke et al. 2001), and this has also been done for other plant species such as *N. attenuata* (Kessler and Baldwin 2001; 2004; Kessler et al. 2004; Voelckel and Baldwin 2004).

Three major signal transduction pathways are known to be involved in the induction of plant volatiles: the octadecanoid, the salicylic-acid and the ethylene pathways (Dicke and Van Poecke 2002; Kessler and Baldwin 2002, for review). Well-characterized genotypes that are altered in these signal transduction pathways are available for *Arabidopsis* (Pieterse and Van Loon 1999; Reymond et al. 2000; Walling 2000). These genotypes allow the analysis of the involvement of the signal transduction pathways with chirurgic accuracy. A single gene has been modified and thus a single step in signal production or signal perception has been altered. These genotypes have been successfully used in the study of induced resistance against phytopathogenic micro-organisms (Pieterse and Van Loon 1999; Walling 2000). These and other well-characterized *Arabidopsis* genotypes are available to investigate the effect of single traits on interactions mediated by herbivore-induced plant volatiles (Van Poecke and Dicke 2002). This will allow to evaluate the new information in the context of induced responses to other environmental variation such as the attack by pathogens (e.g., De Vos et al. 2005; Pieterse and Van Loon 1999).

The major advantages of using *Arabidopsis* for a molecular-ecological approach are that its genome has been sequenced, that a multitude of well-characterized mutants and transgenics is available and that full-genome microarrays are available that can be used to investigate global transcriptome changes in response to biotic interactions (e.g., De Vos et al. 2005; Reymond et al. 2004; Schenk et al. 2000). Moreover, some of these methodologies or the results of their use with *Arabidopsis* may be transferred to other Brassicaceous plants (e.g., Lee et al. 2004). Therefore, information obtained for *Arabidopsis* may be exploited to develop novel approaches for understanding chemical ecology and community ecology of *Brassica*–insect interactions. This will provide important complementary knowledge on the ecology of *Brassica*–insect interactions as obtained through classical methods (e.g., Geervliet et al. 2000; Harvey et al. 2003; Mattiacci et al. 1995; Shiojiri et al. 2001).

For *Arabidopsis* the connection between transcriptomics and proteomics has been made for several biological contexts (Hirai et al. 2005; Peck 2005). Moreover, the connection between transcriptomics and metabolomics has been made (D'Auria and Gershenzon 2005), also in the context of infochemicals (Mercke et al. 2004). When the connection between gene activity and metabolomics has been made, novel tools will be available to tackle many of the questions that have been addressed for many decades in the ecology of insect–plant interactions, i.e., understanding the function of so-called secondary metabolites (Berenbaum et al. 1989; Fraenkel 1959;

Poppy 1999; Schoonhoven et al. in press). Gaining knowledge of mechanisms in terms of transcriptomics, proteomics and metabolomics provides a solid basis for understanding the expression of phenotypes of an organism under different conditions, also termed 'phenomics' (Edwards and Batley 2004; Kahraman et al. 2005). The field of phenomics addresses the documentation of a phenotypic characteristic to a gene and is so far characterized by a deterministic nature where one gene has a single phenotype. However, ecologists are well aware that phenotypes may change with conditions and as such the input of ecologists in phenomics is badly wanted. It may seem fashionable to invent ever new 'omics' for every new and higher layer of integration. However, at the next level of integration, i.e., understanding the total of interactions within a community, we do not need a new 'omics' term: the term community ecology sufficiently covers this area. Linking 'omics' to community ecology is an exciting challenge because it implies linking two very different ways of looking at biological phenomena, viz., a deterministic and highly technology-driven approach and a stochastic and concept-driven approach. First developments in this area show that linking 'omics' with community ecology can be highly rewarding and is likely to answer questions that were difficult to answer so far (Dicke et al. 2004; Howe and Brunner 2005; Kessler et al. 2004; Shimizu and Purugganan 2005). After all, a major challenge for ecologists has been to understand how individual traits of organisms affect species interactions and community dynamics. Breakthroughs in the 'omics' fields provide ecologists with exciting tools to address this through an ecogenomics approach. This allows the most delicate manipulative studies that one can think of, in which mechanistic knowledge of well-characterized genotypes and phenotypic plasticity can be exploited to study the effect of individual plant traits on interactions in ecosystems.

ACKNOWLEDGEMENTS

This work is supported by a VICI grant from the Netherlands Organisation for Scientific Research, NWO (865.03.002) and the European Commission (contract MC-RTN-CT-2003–504720 'ISONET').

REFERENCES

Abrams, P.A., Menge, B.A. and Mittelbach, G.G., 1996. The role of indirect effects in food webs. *In:* Polis, G.A. and Winemiller, K.O. eds. *Food webs: integration of patterns and dynamics*. Chapman & Hall, New York, 371-395.

Agrawal, A.A., 2001. Phenotypic plasticity in the interactions and evolution of species. *Science,* 294 (5541), 321-326.

Ament, K., Kant, M.R., Sabelis, M.W., et al. 2004. Jasmonic acid is a key regulator of spider mite-induced volatile terpenoid and methyl salicylate emission in tomato. *Plant Physiology,* 135 (4), 2025-2037.

Baldwin, I.T., 2001. An ecologically motivated analysis of plant-herbivore interactions in native tobacco. *Plant Physiology,* 127 (4), 1449-1458.

Baldwin, I.T., Halitschke, R., Kessler, A., et al. 2001. Merging molecular and ecological approaches in plant-insect interactions. *Current Opinion in Plant Biology,* 4 (4), 351-358.

Bartelt, R.J., Schaner, A.M. and Jackson, L.L., 1985. Cis-vaccenyl acetate as an aggregation pheromone in *Drosophila melanogaster*. *Journal of Chemical Ecology,* 11 (12), 1747-1756.
Berenbaum, M.R., Zangerl, A.R. and Lee, K., 1989. Chemical barriers to adaptation by a specialist herbivore. *Oecologia,* 80 (4), 501-506.
Bernays, E.A., 1998. Evolution of feeding behavior in insect herbivores: success seen as different ways to eat without being eaten. *Bioscience,* 48 (1), 35-44.
Bonsall, M.B. and Hassell, M.P., 1997. Apparent competition structures ecological assemblages. *Nature,* 388 (6640), 371-373.
Bouwmeester, H.J., Kappers, I.F., Verstappen, F.W.A., et al. 2003. Exploring multi-trophic plant-herbivore interactions for new crop protection methods. *In: International congress of crop science and technology, 10-12 November 2003, Glasgow. Vol. 2,* Vol. 2. British Crop Protection Council, Alton, 1123-1134. [http://library.wur.nl/wasp/bestanden/LUWPUBRD_00325287_A502_001.pdf]
D'Auria, J.C. and Gershenzon, J., 2005. The secondary metabolism of *Arabidopsis thaliana*: growing like a weed. *Current Opinion in Plant Biology,* 8 (3), 308-316.
De Vos, M., Van Oosten, V.R., Van Poecke, R.M.P., et al. 2005. Signal signature and transcriptome changes of *Arabidopsis* during pathogen and insect attack. *Molecular Plant Microbe Interactions,* 18 (9), 923-937.
Dicke, M., 1999. Evolution of induced indirect defence of plants. *In:* Tollrian, R. and Harvell, C.D. eds. *The ecology and evolution of inducible defenses*. Princeton University Press, Princeton, 62-88.
Dicke, M., 2000. Chemical ecology of host-plant selection by herbivorous arthropods: a multitrophic perspective. *Biochemical Systematics and Ecology,* 28 (7), 601-617.
Dicke, M. and Bruin, J., 2001. Chemical information transfer between plants: back to the future. *Biochemical Systematics and Ecology,* 29 (10), 981-994.
Dicke, M. and Grostal, P., 2001. Chemical detection of natural enemies by arthropods: an ecological perspective. *Annual Review of Ecology and Systematics,* 32, 1-23.
Dicke, M. and Hilker, M., 2003. Induced plant defences: from molecular biology to evolutionary ecology. *Basic and Applied Ecology,* 4 (1), 3-14.
Dicke, M. and Sabelis, M.W., 1992. Costs and benefits of chemical information conveyance: proximate and ultimate factors. *In:* Roitberg, B.D. and Isman, M.B. eds. *Insect chemical ecology: an evolutionary approach*. Chapman & Hall, New York, 122-155.
Dicke, M. and Takken, W. (eds.), in press. *Chemical ecology: from gene to ecosystem*. Springer, Dordrecht. Wageningen UR Frontis Series no. 16. [http://library.wur.nl/frontis/chemical_ecology/]
Dicke, M., Van Beek, T.A., Posthumus, M.A., et al. 1990. Isolation and identification of volatile kairomone that affects acarine predator-prey interactions: involvement of host plant in its production. *Journal of Chemical Ecology,* 16 (2), 381-396.
Dicke, M. and Van Loon, J.J.A., 2000. Multitrophic effects of herbivore-induced plant volatiles in an evolutionary context. *Entomologia Experimentalis et Applicata,* 97 (3), 237-249.
Dicke, M., Van Loon, J.J.A. and De Jong, P.W., 2004. Ecogenomics benefits community ecology. *Science,* 305 (5684), 618-619.
Dicke, M. and Van Poecke, R.M.P., 2002. Signalling in plant-insect interactions: signal transduction in direct and indirect plant defence. *In:* Scheel, D. and Wasternack, C. eds. *Plant signal transduction.* Oxford University Press, Oxford, 289-316. Frontiers in Molecular Biology no. 38.
Dicke, M., Van Poecke, R.M.P. and De Boer, J.G., 2003. Inducible indirect defence of plants: from mechanisms to ecological functions. *Basic and Applied Ecology,* 4 (1), 27-42.
Dicke, M. and Vet, L.E.M., 1999. Plant-carnivore interactions: evolutionary and ecological consequences for plant, herbivore and carnivore. *In:* Olff, H., Brown, V.K. and Drent, R.H. eds. *Herbivores between plants and predators: the 38th symposium of the British Ecological Society in cooperation with the Netherlands Ecological Society, held at Wageningen Agricultural University, The Netherlands, 1997.* Blackwell Science, Oxford, 483-520.
Edwards, D. and Batley, J., 2004. Plant bioinformatics: from genome to phenome. *Trends in Biotechnology,* 22 (5), 232-237.
Fatouros, N.E., Huigens, M.E., Van Loon, J.J.A., et al. 2005. Chemical communication: butterfly anti-aphrodisiac lures parasitic wasps. *Nature,* 433 (7027), 704.
Fiehn, O., 2002. Metabolomics: the link between genotypes and phenotypes. *Plant Molecular Biology,* 48 (1/2), 155-175.
Fitzpatrick, M.J., Ben-Shahar, Y., Smid, H.M., et al. 2005. Candidate genes for behavioural ecology. *Trends in Ecology and Evolution,* 20 (2), 96-104.

Fraenkel, G.S., 1959. The raison d'être of secondary plant substances. *Science,* 129 (3361), 1466-1470.

Geervliet, J.B.F., Verdel, M.S.W., Snellen, H., et al. 2000. Coexistence and niche segregation by field populations of the parasitoids *Cotesia glomerata* and *C. rubecula* in the Netherlands: predicting field performance from laboratory data. *Oecologia,* 124 (1), 55-63.

Hamilton, A.J. and Baulcombe, D.C., 1999. A species of small antisense RNA in posttranscriptional gene silencing in plants. *Science,* 286 (5441), 950-952.

Harvey, J.A., Van Dam, N.M. and Gols, R., 2003. Interactions over four trophic levels: foodplant quality affects development of a hyperparasitoid as mediated through a herbivore and its primary parasitoid. *Journal of Animal Ecology,* 72 (3), 520-531.

Hedlund, K., Vet, L.E.M. and Dicke, M., 1996. Generalist and specialist parasitoid strategies of using odours of adult drosophilid flies when searching for larval hosts. *Oikos,* 77 (3), 390-398.

Heidel, A.J. and Baldwin, I.T., 2004. Microarray analysis of salicylic acid- and jasmonic acid-signalling in responses of *Nicotiana attenuata* to attack by insects from multiple feeding guilds. *Plant Cell and Environment,* 27 (11), 1362-1373.

Hilker, M. and Meiners, T., 2002. Induction of plant responses to oviposition and feeding by herbivorous arthropods: a comparison. *Entomologia Experimentalis et Applicata,* 104 (1), 181-192.

Hilker, M., Rohfritsch, O. and Meiners, T., 2002. The plant's response towards insect egg depostion. *In:* Hilker, M. and Meiners, T. eds. *Chemoecology of insect eggs and egg deposition.* Blackwell Publishers, Berlin, 205-233.

Hirai, M.Y., Klein, M., Fujikawa, Y., et al. 2005. Elucidation of gene-to-gene and metabolite-to-gene networks in *Arabidopsis* by integration of metabolomics and transcriptomics. *Journal of Biological Chemistry,* 280 (27), 25590-25595.

Howe, G.T. and Brunner, A.M., 2005. An evolving approach to understanding plant adaptation. *New Phytologist,* 167 (1), 1-5.

Hunter, M.D., 2002. A breath of fresh air: beyond laboratory studies of plant volatile-natural enemy interactions. *Agricultural and Forest Entomology,* 4 (2), 81-86.

Jacobs, S.P., Liggins, A.P., Zhou, J.J., et al. 2005. OS-D-like genes and their expression in aphids (Hemiptera: Aphididae). *Insect Molecular Biology,* 14 (4), 423-432.

Kahraman, A., Avramov, A., Nashev, L.G., et al. 2005. PhenomicDB: a multi-species genotype/phenotype database for comparative phenomics. *Bioinformatics,* 21 (3), 418-420.

Kats, L.B. and Dill, L.M., 1998. The scent of death: chemosensory assessment of predation risk by prey animals. *Ecoscience,* 5 (3), 361-394.

Kessler, A., in press. Plant-insect interactions in the era of consolidation in biological sciences: *Nicotiana attenuata* as an ecological expression system. *In:* Dicke, M. and Takken, W. eds. *Chemical ecology: from gene to ecosystem.* Springer, Dordrecht. Wageningen UR Frontis Series no. 16. [http://library.wur.nl/frontis/chemical_ecology/03_kessler.pdf]

Kessler, A. and Baldwin, I.T., 2001. Defensive function of herbivore-induced plant volatile emissions in nature. *Science,* 291 (5511), 2141-2144.

Kessler, A. and Baldwin, I.T., 2002. Plant responses to insect herbivory: the emerging molecular analysis. *Annual Review of Plant Biology,* 53, 299-328.

Kessler, A. and Baldwin, I.T., 2004. Herbivore-induced plant vaccination. Part I. The orchestration of plant defenses in nature and their fitness consequences in the wild tobacco *Nicotiana attenuata. Plant Journal,* 38 (4), 639-649.

Kessler, A., Halitschke, R. and Baldwin, I.T., 2004. Silencing the jasmonate cascade: induced plant defenses and insect populations. *Science,* 305 (5684), 665-668.

Lee, H.S., Wang, J.L., Tian, L., et al. 2004. Sensitivity of 70-mer oligonucleotides and cDNAs for microarray analysis of gene expression in *Arabidopsis* and its related species. *Plant Biotechnology Journal,* 2 (1), 45-57.

Mattiacci, L., Dicke, M. and Posthumus, M.A., 1995. Beta-glucosidase: an elicitor of herbivore-induced plant odor that attracts host-searching parasitic wasps. *Proceedings of the National Academy of Sciences of the United States of America,* 92 (6), 2036-2040.

Mercke, P., Kappers, I.F., Verstappen, F.W.A., et al., 2004. Combined transcript and metabolite analysis reveals genes involved in spider mite induced volatile formation in cucumber plants. *Plant Physiology,* 135 (4), 2012-2024.

Mitchell-Olds, T., 2001. *Arabidopsis thaliana* and its wild relatives: a model system for ecology and evolution. *Trends in Ecology and Evolution,* 16 (12), 693-700.

Morin, P.J., 1999. *Community ecology.* Blackwell, Malden.

Omacini, M., Chaneton, E.J., Ghersa, C.M., et al. 2001. Symbiotic fungal endophytes control insect host-parasite interaction webs. *Nature,* 409 (6816), 78-81.

Papaj, D.R. and Lewis, A.C. (eds.), 1993. *Insect learning: ecological and evolutionary perspectives.* Chapman & Hall, New York.

Peck, S.C., 2005. Update on proteomics in *Arabidopsis*: where do we go from here? *Plant Physiology,* 138 (2), 591-599.

Penn, D.J., 2002. The scent of genetic compatibility: sexual selection and the major histocompatibility complex. *Ethology,* 108 (1), 1-21.

Pieterse, C.M.J. and Van Loon, L.C., 1999. Salicylic acid-independent plant defence pathways. *Trends in Plant Science,* 4 (2), 52-58.

Poppy, G.M., 1999. The raison d'être of secondary plant chemicals? *Trends in Plant Science,* 4 (3), 82-83.

Rasmann, S., Kollner, T.G., Degenhardt, J., et al. 2005. Recruitment of entomopathogenic nematodes by insect-damaged maize roots. *Nature,* 434 (7034), 732-737.

Reymond, P., Bodenhausen, N., Van Poecke, R.M.P., et al. 2004. A conserved transcript pattern in response to a specialist and a generalist herbivore. *Plant Cell,* 16 (11), 3132-3147.

Reymond, P., Weber, H., Damond, M., et al. 2000. Differential gene expression in response to mechanical wounding and insect feeding in *Arabidopsis. Plant Cell,* 12 (5), 707-719.

Robertson, D., 2004. VIGS vectors for gene silencing: many targets, many tools. *Annual Review of Plant Biology,* 55, 495-519.

Roda, A.L. and Baldwin, I.T., 2003. Molecular technology reveals how the induced direct defenses of plants work. *Basic and Applied Ecology,* 4 (1), 15-26.

Roitberg, B.D. and Isman, M.B. (eds.), 1992. *Insect chemical ecology: an evolutionary approach.* Chapman & Hall, New York.

Rosenthal, G.A. and Berenbaum, M.R. (eds.), 1992. *Herbivores: their interaction with secondary plant metabolites.* 2nd edn. Academic Press, New York.

Rott, A.S. and Godfray, H.C.J., 2000. The structure of a leafminer-parasitoid community. *Journal of Animal Ecology,* 69 (2), 274-289.

Sabelis, M.W., Van Baalen, M., Bakker, F.M., et al. 1999. The evolution of direct and indirect plant defence against herbivorous arthropods. *In:* Olff, H., Brown, V.K. and Drent, R.H. eds. *Herbivores between plants and predators: the 38th symposium of the British Ecological Society in cooperation with the Netherlands Ecological Society, held at Wageningen Agricultural University, The Netherlands, 1997.* Blackwell Science, Oxford, 109-166.

Schenk, P.M., Kazan, K., Wilson, I., et al. 2000. Coordinated plant defense responses in *Arabidopsis* revealed by microarray analysis. *Proceedings of the National Academy of Sciences of the United States of America,* 97 (21), 11655-11660.

Schoonhoven, L.M., Van Loon, J.J.A. and Dicke, M., in press. *Insect-plant biology.* 2nd edn. Oxford University Press, Oxford.

Shimizu, K.K. and Purugganan, M.D., 2005. Evolutionary and ecological genomics of arabidopsis. *Plant Physiology,* 138 (2), 578-584.

Shiojiri, K., Takabayashi, J., Yano, S., et al. 2001. Infochemically mediated tritrophic interaction webs on cabbage plants. *Population Ecology,* 43 (1), 23-29.

Stowe, M.K., Turlings, T.C., Loughrin, J.H., et al. 1995. The chemistry of eavesdropping, alarm, and deceit. *Proceedings of the National Academy of Sciences of the United States of America,* 92 (1), 23-28.

Thaler, J.S., 1999. Jasmonate-inducible plant defences cause increased parasitism of herbivores. *Nature,* 399 (6737), 686-688.

Thaler, J.S., 2002. Effect of jasmonate-induced plant responses on the natural enemies of herbivores. *Journal of Animal Ecology,* 71 (1), 141-150.

Thaler, J.S., Farag, M.A., Pare, P.W., et al. 2002. Jasmonate-deficient plants have reduced direct and indirect defences against herbivores. *Ecology Letters,* 5 (6), 764-774.

Thaler, J.S., Stout, M.J., Karban, R., et al. 2001. Jasmonate-mediated induced plant resistance affects a community of herbivores. *Ecological Entomology,* 26 (3), 312-324.

Tollrian, R. and Harvell, C.D. (eds.), 1999. *The ecology and evolution of inducible defenses.* Princeton University Press, Princeton.

Turlings, T.C.J. and Benrey, B., 1998. Effects of plant metabolites on the behavior and development of parasitic wasps. *Ecoscience,* 5 (3), 321-333.

Turlings, T.C.J., Wäckers, F.L., Vet, L.E.M., et al. 1993. Learning of host-finding cues by hymenopterous parasitoids. *In:* Papaj, D.R. and Lewis, A.C. eds. *Insect learning: ecological and evolutionary perspectives*. Chapman & Hall, New York, 51-79.

Van den Boom, C.E.M., Van Beek, T.A., Posthumus, M.A., et al. 2004. Qualitative and quantitative variation among volatile profiles induced by *Tetranychus urticae* feeding on plants from various families. *Journal of Chemical Ecology,* 30 (1), 69-89.

Van Donk, E., in press. Food-web interactions in lakes: what is the impact of chemical information conveyance? *In:* Dicke, M. and Takken, W. eds. *Chemical ecology: from gene to ecosystem.* Springer, Dordrecht. Wageningen UR Frontis Series no. 16. [http://library.wur.nl/frontis/chemical_ecology/10_van_donk.pdf]

Van Poecke, R.M.P. and Dicke, M., 2002. Induced parasitoid attraction by *Arabidopsis thaliana*: involvement of the octadecanoid and the salicylic acid pathway. *Journal of Experimental Botany,* 53 (375), 1793-1799.

Van Poecke, R.M.P. and Dicke, M., 2003. Signal transduction downstream of salicylic and jasmonic acid in herbivory-induced parasitoid attraction by *Arabidopsis* is independent of JAR1 and NPR1. *Plant Cell and Environment,* 26 (9), 1541-1548.

Van Poecke, R.M.P. and Dicke, M., 2004. Indirect defence of plants against herbivores: using *Arabidopsis thaliana* as a model plant. *Plant Biology,* 6 (4), 387-401.

Van Poecke, R.M.P., Posthumus, M.A. and Dicke, M., 2001. Herbivore-induced volatile production by *Arabidopsis thaliana* leads to attraction of the parasitoid *Cotesia rubecula*: chemical, behavioral, and gene-expression analysis. *Journal of Chemical Ecology,* 27 (10), 1911-1928.

Van Tol, R.W.H.M., Van der Sommen, A.T.C., Boff, M.I.C., et al. 2001. Plants protect their roots by alerting the enemies of grubs. *Ecology Letters,* 4 (4), 292-294.

Vet, L.E.M., 1999. From chemical to population ecology: infochemical use in an evolutionary context. *Journal of Chemical Ecology,* 25 (1), 31-49.

Voelckel, C. and Baldwin, I.T., 2004. Herbivore-induced plant vaccination. Part II. Array-studies reveal the transience of herbivore-specific transcriptional imprints and a distinct imprint from stress combinations. *Plant Journal,* 38 (4), 650-663.

Walling, L.L., 2000. The myriad plant responses to herbivores. *Journal of Plant Growth Regulation,* 19 (2), 195-216.

Wertheim, B., 2005. Evolutionary ecology of communication signals that induce aggregative behaviour. *Oikos,* 109 (1), 117-124.

Wertheim, B., Allemand, R., Vet, L.E.M., et al. in press. Effects of aggregation pheromone on individual behaviour and food web interactions: a field study on *Drosophila. Ecological Entomology.*

Wertheim, B., Van Baalen, E.J., Dicke, M., et al. 2005. Pheromone-mediated aggregation in nonsocial arthropods: an evolutionary ecological perspective. *Annual Review of Entomology,* 50, 321-346.

Wageningen UR Frontis Series

1. A.G.J. Velthuis, L.J. Unnevehr, H. Hogeveen and R.B.M. Huirne (eds.): *New Approaches to Food-Safety Economics*. 2003
ISBN 1-4020-1425-2; Pb: 1-4020-1426-0
2. W. Takken and T.W. Scott (eds.): *Ecological Aspects for Application of Genetically Modified Mosquitoes*. 2003
ISBN 1-4020-1584-4; Pb: 1-4020-1585-2
3. M.A.J.S. van Boekel, A. Stein and A.H.C. van Bruggen (eds.): *Proceedings of the Frontis workshop on Bayesian Statistics and quality modelling*. 2003
ISBN 1-4020-1916-5
4. R.H.G. Jongman (ed.): *The New Dimensions of the European Landscape*. 2004
ISBN 1-4020-2909-8; Pb: 1-4020-2910-1
5. M.J.J.A.A. Korthals and R.J.Bogers (eds.): *Ethics for Life Scientists*. 2004
ISBN 1-4020-3178-5; Pb: 1-4020-3179-3
6. R.A. Feddes, G.H.de Rooij and J.C. van Dam (eds.): *Unsaturated-zone modeling*. Progress, challenges and applications. 2004 ISBN 1-4020-2919-5
7. J.H.H. Wesseler (ed.): *Environmental Costs and Benefits of Transgenic Crops*. 2005 ISBN 1-4020-3247-1; Pb: 1-4020-3248-X
8. R.S. Schrijver and G. Koch (eds.): *Avian Influenza*. Prevention and Control. 2005
ISBN 1-4020-3439-3; Pb: 1-4020-3440-7
9. W. Takken, P. Martens and R.J. Bogers (eds.): *Environmental Change and Malaria Risk*. Global and Local Implications. 2005
ISBN 1-4020-3927-1; Pb: 1-4020-3928-X
10. L.J.E.J. Gilissen, H.J. Wichers, H.F.J. Savelkoul and R.J. Bogers, (eds.): *Allergy Matters*. New Approaches to Allergy Prevention and Management. 2005
ISBN 1-4020-3895-X; Pb: 1-4020-3896-8
11. B.G.J. Knols and C. Louis (eds.): *Briding Laboratory and Field Research for Genetic Control of Disease Vectors*. 2006
ISBN 1-4020-3800-3; Pb: 1-4020-3799 6
12. B. Tress, G. Tress, G. Fry and P. Opdam (eds.): *From Landscape Research to Landscape Planning*. Aspects of Integration, Education and Application. 2006
ISBN 1-4020-3979-4; Pb: 1-4020-3978-6
13. J. Hassink and M. van Dijk (eds.): *Farming for Health*. Green-Care Farming Across Europe and the United States of America. 2006
ISBN 1-4020-4540-9; Pb: 1-4020-4541-7
14. R. Ruben, M. Slingerland and H. Nijhoff (eds.): *The Agro-Food Chains and Networks for Development*. 2006 ISBN 1-4020-4592-1; Pb: 1-4020-4600-6
15. C.J.M. Ondersteijn, J.H.M. Wijnands, R.B.M. Huiren and O. van Kooten (eds.): *Quantifying the Agri-Food Supply Chain*.
ISBN 1-4020-4692-8; Pb: 1-4020-4693-6
16. M. Dicke and W. Takken (eds.): *Chemical Ecology*. From Gene to Ecosystem. 2006 ISBN 1-4020-4783-5; Pb: 1-4020-4792-4

springer.com

Printed by Printforce, the Netherlands